Induction Motors - Latest Research and Applications

Edited by Adel El-Shahat

Published in London, United Kingdom

Induction Motors - Latest Research and Applications
http://dx.doi.org/10.5772/intechopen.1003548
Edited by Adel El-Shahat

Contributors
Abdellah Oukassi, Allan R.A. Manito, Eduardo Cabal-Yepez, Efrain Ramirez-Velasco, Elielson F.S. Barbosa, Emrah Zerdali, Francisco Javier Villalobos-Pina, Jhennifer F. dos Santos, Josué A. Reyes-Malanche, José Rodríguez, Lucas H.B. Santos, Marcelo O. Silva, Marco Rivera, Mohaned Menzou, Patrick Wheeler, Saad Khadar, Tahar Abbassi, Wellington S. Fonseca, Yasmim M.D. Lisboa

First published in London, United Kingdom, 2025 by IntechOpen
IntechOpen is the global imprint of INTECHOPEN LIMITED, registered in England and Wales, registration number: 11086078, 167-169 Great Portland Street, London, W1W 5PF, United Kingdom

For EU product safety concerns: IN TECH d.o.o., Prolaz Marije Krucifikse Kozulić 3, 51000 Rijeka, Croatia, info@intechopen.com or visit our website at intechopen.com.

British Library Cataloguing-in-Publication Data
A catalogue record for this book is available from the British Library

Induction Motors - Latest Research and Applications
Edited by Adel El-Shahat
p. cm.
Print ISBN 978-0-85014-811-4
Online ISBN 978-0-85014-810-7
eBook (PDF) ISBN 978-0-85014-812-1

If disposing of this product, please recycle the paper responsibly.

Meet the editor

Dr. Adel El-Shahat is an Assistant Professor of Energy Technology at the School of Engineering Technology, Purdue University, USA. He is the founder and director of Advanced Power Units and Renewable Distributed Energy Lab (A_PURDUE). His research focuses on modeling, design, optimization, simulation, analysis, and control of renewable energy systems, smart nano and microgrids, electric mobility and transportation electrification, wireless charging of electric vehicles, climate change, electric vehicles, special-purpose electric machines, deep learning techniques, distributed generation systems, thermoelectric generation, special power electronics converters, power systems, energy storage and conservation, and engineering education. To his credit, Dr. El-Shahat has 11 books, 7 book chapters, 85 journal papers, 78 conference papers, and 114 other publications. Dr. El-Shahat is a Senior Member of IEEE, a Senior Member of the UACEE (Institute of Research Engineers and Doctors), a member of AAAS (American Association for the Advancement of Science), the U. S. National Committee of CIGRE, the ASEE (American Society for Engineering Education), the IAENG (International Association of Engineers), the IACSIT (Int. Association of Computer Science and Information Technology), the Arab Research Institute for Science & Engineering, as well as a member of the Scientific and Technical Committee & Editorial Review Board on Natural and Applied Sciences, (World Academy of Science Engineering and Technology) WASET.

Contents

Preface

This book, *Induction Motors – Latest Research and Applications* is a collection of reviewed and relevant research chapters concerning the developments within the electrical and electronic engineering field of study. The book includes scholarly contributions by various authors and is edited by a group of experts pertinent to this field. Each contribution comes as a separate chapter, complete in itself, but directly related to the book's topics and objectives.

The book includes chapters:

1. Motor Monitoring and Analysis System (SMAM) as a Resource for Predictive Maintenance in Industrial Environments through Digital Twin and IIoT

2. Open Circuit Fault Diagnosis in Induction Motor Driver Inverter

3. Implementing ADRC and PI Controls for a Sensorless Induction Motor Supplied by a Multi-Level Converter and Using PV Sources

4. A Full-Order Terminal Sliding-Mode Observer-Based Optimal Resistance Estimation for High-Performance Control Applications

5. Predictive Control for Induction Machines

The target audience comprises scholars and specialists in the field.

Adel El-Shahat
Purdue University West Lafayette,
Indiana, United States of America

Chapter 1

Motor Monitoring and Analysis System (SMAM) as a Resource for Predictive Maintenance in Industrial Environments through Digital Twin and IIoT

Jhennifer F. dos Santos, Yasmim M.D. Lisboa, Elielson F.S. Barbosa, Lucas H.B. Santos, Allan R.A. Manito, Marcelo O. Silva and Wellington S. Fonseca

Abstract

The Motor Monitoring and Analysis System (SMAM) was developed as a tool that can help with the predictive maintenance of electric motors through Digital Twin (DT) and the Industrial Internet of Things (IIoT). To do this, SMAM has an embedded system made up of the ESP32 microcontroller and temperature, voltage and current sensors, in which the measurements taken are stored in the Firebase web database. It is worth noting that the sensors are calibrated using the least squares and linear regression mathematical methods, as well as using the ET-5062 energy analyzer as a reference to ensure that the measurement is reliable. In addition, strong coupling simulations are carried out in the Finite Element Method Magnetics (FEMM) software, which makes it possible to visualize the magnetic flux and the internal temperature of the motor by entering the measured current data. In this way, SMAM is a promising resource for the manufacturing environment.

Keywords: induction motors, finite element analysis, industrial internet of things, digital twin, maintenance

1. Introduction

Electric motors have a significant presence in the industry, acting as essential components in various production processes [1]. This equipment is considered a critical asset, as it drives machines and systems that are fundamental to the continuous and efficient operation of factories. Because of its importance, the failure of an electric motor can result in unplanned downtime, loss of productivity, and high maintenance costs [2]. For this reason, constant monitoring of electric motors is crucial to ensure good operating performance.

In recent years, researchers have been exploring the development of tools for monitoring induction motors. One example is the article [3] that proposes an automated induction motor monitoring system with IoT, using a Raspberry Pi 4 and the Blynk IoT interface to protect the motor in abnormal conditions, such as overvoltage and high temperature, ensuring continuous operation or automatic shutdown. The study carried out by Patel et al. [4] monitors the condition of three-phase induction motors by analyzing the stator current and vibration, using instruments such as the Digital Storage Oscilloscope and the Vibration Analyzer to diagnose faults with the support of MATLAB.

In addition, Noyjeen et al. [5] propose the control and monitoring of three-phase motors with IoT, using the Raspberry Pi to collect data from sensors and send it to a server, where it is processed to detect faults in components such as the rotor and stator. The work by Noyjeen et al. [5] presents a system for monitoring three-phase motor parameters using the ESP32 microcontroller, with cloud storage in ThingSpeak and notifications *via* the Line Notify API, allowing remote access and real-time alerts when performance limits are exceeded. This research reinforces the relevance of innovative solutions for the efficient monitoring of electric motors in industry.

Within the context of these advances, the Industrial Internet of Things (IIoT) has emerged as a key technology. The IIoT, which is an extension of the Internet of Things (IoT), is specifically focused on the industrial environment [6, 7]. While IoT covers the connectivity and automation of devices in a wide range of applications, IIoT is aimed at optimizing industrial processes, connecting sensors, machines, and systems to collect and analyze data in real time. The application of IIoT in the industrial sector brings significant advantages, such as the possibility of carrying out predictive maintenance, avoiding unexpected failures, and extending the useful life of assets [8].

Another emerging technology that is being widely adopted in the industry is the Digital Twin. A Digital Twin is a virtual representation of a physical asset, which allows you to simulate and monitor its behavior in real time [9]. There are different types of Digital Twin, including the Product, Process and Performance Digital Twin, each focused on different aspects of an asset's operation [10]. In industry, Digital Twins have been used to monitor engines, providing a detailed view of their operation and allowing simulations that predict future behavior based on real data [11]. This approach allows for more precise control and a rapid response to any anomalies detected, increasing the reliability and efficiency of electric motors.

In this context, this article will present the optimization of the acquisition module that makes up the Motor Monitoring and Analysis System (from Portuguese, Sistema de Monitoramento e Análise de Motores - SMAM) proposed by Santos et al. [12]. This system integrates IIoT and Digital Twin technologies as a tool that can help with the predictive maintenance of electric motors. SMAM uses an embedded system with an ESP32 microcontroller and sensors to measure critical variables such as temperature, voltage, and current, whose data is stored in a web database. In addition, advanced simulations are carried out to visualize the internal behavior of the engine and these results are displayed on a web panel, making SMAM a promising tool for the industry.

The reason for building a new acquisition module is that the work presented by Santos et al. [12] had a significant error in one of the measurements of the current sensor that makes up the system and which plays a crucial role in the Digital Twin model. The current values measured by the module are incremented in the simulation carried out in the Finite Element Method Magnetics (FEMM) software. Therefore, in addition to the development of a new module, the contribution of this work is the application of a calibration methodology using the least squares method and linear regression.

2. The motor monitoring and analysis system

The Motor Monitoring and Analysis System (SMAM) consists of an IIoT module that measures the temperature, voltage, and phase current of induction motors using non-invasive sensors. The module is equipped with the ESP32 microcontroller, which transmits measurements in real time to the cloud *via* a Wi-Fi connection. In addition, computer simulations of the strong coupling of the monitored motor are carried out. **Figure 1** illustrates the general layout of the SMAM.

As described above, this work proposes the optimization of the embedded system developed by Santos et al. [12]. Initially, a module was built on a perforated phenolite plate measuring 12 cm long by 9.3 cm wide, which accommodates the ESP32 microcontroller and the voltage, current, and temperature sensors, as well as providing space for the inclusion of new sensors if necessary, as shown in **Figure 2**.

2.1 Acquisition module

The optimized module is managed by the ESP32-WROOM-32 microcontroller, which is the backbone of the monitoring since it is responsible for receiving the sensor measurements and sending them to a cloud database. The ESP32 sends this information *via* the IEEE 802.11 protocol [13].

The ESP32 microcontroller has 32-bit dual-core processors operating at 240 MHz, analog-to-digital converters, and an internal antenna with a power of 20 dBm, supporting a data rate of up to 150 Mbps. It also has a sampling rate of 1 kHz, low bit error rate (BER), and performs high-speed floating point operations. This device has gained popularity in the industry due to its capabilities, being used, for example, in the WCD-GT100-ETH and WCD-GT100 gateways from WEG [14].

It is worth noting that the purpose of this work is restricted to optimizing the existing module, so it was not necessary to replace the sensors and microcontroller used in the [12] work, and the module's wiring diagram 3 was maintained.

The acquisition module consists of three current sensors model SCT-013 (**Figure 3b**), which are current transformers capable of measuring alternating current (AC) of up to 100 A [15]. These sensors operate on the principles of electromagnetic

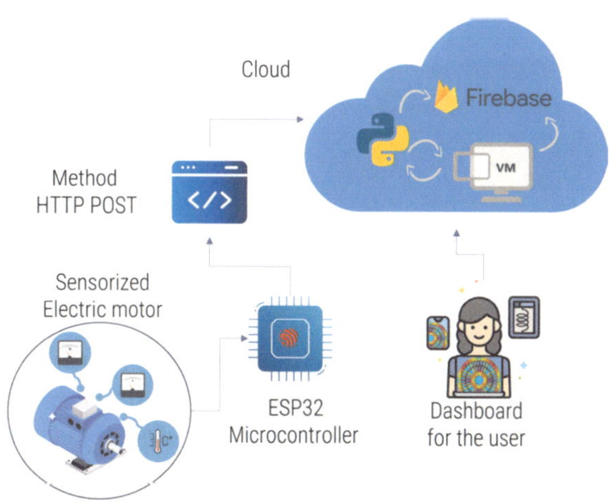

Figure 1.
The functioning of the SMAM proposed by Santos et al. [12].

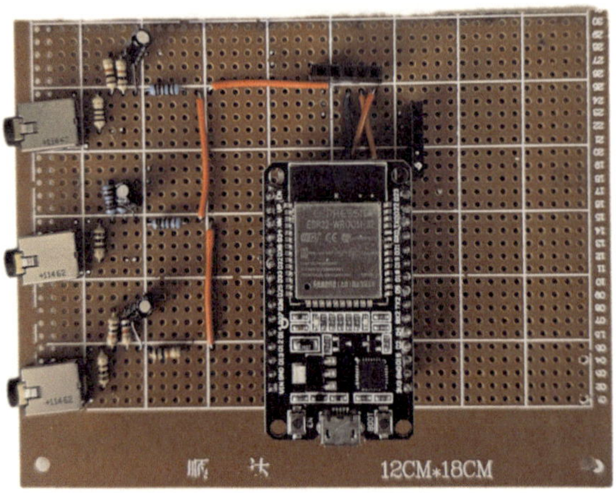

Figure 2.
New acquisition module built for SMAM.

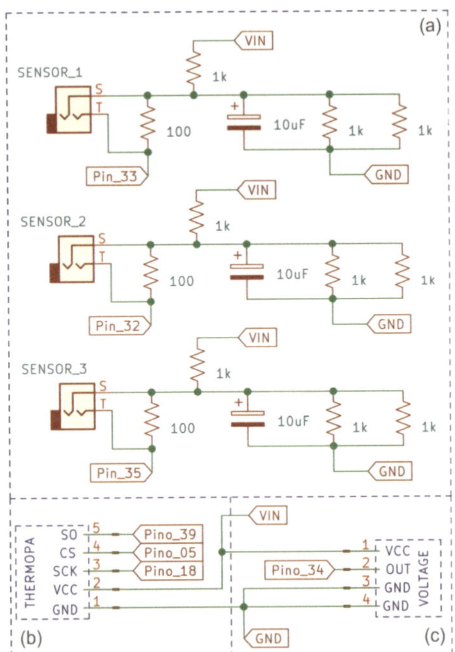

Figure 3.
Electrical schematic of the module [12].

induction, where the magnetic field generated by the alternating current in the primary conductor induces a proportional current in the sensor's secondary circuit. Each SCT-013 is connected to one of the motor's phases, allowing precise measurement of the current in each phase.

For voltage measurement, the ZMPT101B module is used (**Figure 4a**). This sensor is capable of measuring alternating voltages (AC) of up to 250 V, with a margin of error of ±1% [16]. The ZMPT101B includes a voltage transformer and an RC circuit with amplifiers, which condition the voltage signal into a form suitable for reading by the ESP32 microcontroller.

For temperature measurement, the system relies on a type K thermocouple, which has a measuring range from 0 to 600°C. The K-type thermocouple operates on the basis of the Seebeck Effect, where the junction of two different metals, in this case, Cromel and Alumel, generates an electrical voltage proportional to the temperature difference between the hot junction and the reference junction. The signal generated by the thermocouple is in millivolts, which makes it necessary to use the MAX6675, a specialized analog-to-digital converter.

The MAX6675 reads the voltage difference generated by the thermocouple and compares it with the voltage of an internal temperature sensor diode, converting these values into an accurate temperature reading. The integration of the type K thermocouple with the MAX6675 allows the system to obtain reliable and accurate temperature measurements. The type K thermocouple and the MAX6675 are shown in **Figure 4c** and **d**, respectively.

2.2 Sensor calibration

Linear regression and least squares statistical methods were used to adjust the measurements taken by the on-board system. Linear regression is a specific application of the least squares method, where the relationship between variables is modeled by a linear equation and the parameters of this equation are estimated using the least squares method.

Linear regression aims to model the relationship between two variables using a linear equation, where X is the independent variable and y is the dependent variable. Eq. 1 illustrates this model, where i represents each of the n observations in the database, a is the intercept of the regression line on the y-axis, b is the slope of the line and ε is the residual representing the values of y not explained by the model [17].

$$y_i = ax_i + b + \varepsilon_i \tag{1}$$

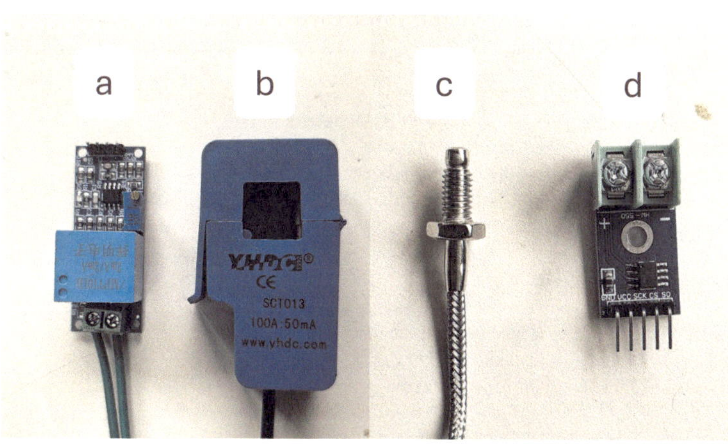

Figure 4.
Voltage (a) and current (b) sensors, as well as the k-type thermocouple (c) and MAX6675 integrated circuit (d) used in the SMAM.

The least squares method is used to estimate the parameters a and b of the linear regression equation. This method seeks to minimize the sum of squares of the residuals (SSE), which are the differences between the observed values of y and the values predicted by the regression equation [18]. With this, it is possible to determine the values of aaa and b that minimize the sum of the squares of the residuals, as shown in Eq. 2.

$$SSE = \sum_{i=1}^{n} \left(y_i - (mx_i + b)\right)^2 \tag{2}$$

To find the SSE, calculate the partial derivatives in relation to a and b and equal them to zero to find the values that minimize the function. Once the values of a and b have been estimated using the least squares method, the linear regression equation can be used to make predictions of y based on the values of x.This results in predictions that represent the best possible estimate of the relationship between the variables, given the distribution of the observed data.

As mentioned earlier, the engine was monitored by the ET-5062 and the optimized SMAM system. Based on this, a Python script was developed to automate the process of calibrating the SMAM's current sensors, using linear regression and least squares, in order to make it easier to identify deviations and improve the accuracy of measurements.

The script used the Pandas library for manipulating and analyzing structured data, allowing data to be loaded from different formats such as CSV, xlsx, and SQL [19]. In addition, the Matplotlib library was used to visualize the data. According to Ref. [20], this library generates high-quality customizable graphics, allowing for intuitive and harmonious data visualization.

The calibration methods were applied using the $sklearn.linear_model$ library, which has the *LinearRegression* function of adjusting the linear regression model to the data using the least squares method [21]. The *fit* method of the *LinearRegression* class calculates the regression coefficients that minimize the sum of the squares of the differences between the actual observations and the model's predictions. In this way, the calculated coefficients represent the parameters of the regression line, solving the least squares problem.

Figure 5 shows the use of *LinearRegression* and the *fit* method for the values of the current sensor that monitors Phase A. It is important to note that these functions were also applied to the motor's Phase B and C sensors.

```
# Upload sensor and commercial sensor and SMAM data
medicao_SMAM = pd.read_excel(medicao_SMAM)
medicao_anali = pd.read_excel(medicao_anali)

# Separate data into sensor readings and corresponding
# measurements for PhaseA
FaseA_SMAM = medicao_SMAM[['Fase A']]

# Initialize and adjust the linear regression model for Phase A
modelo_FaseA = LinearRegression()
modelo_FaseA.fit(FaseA_SMAM, FaseA_anali)

# Regression coefficients for Phase A
coeficiente_FaseA = modelo_FaseA.coef_[0][0]
intercepto_FaseA = modelo_FaseA.intercept_[0]
```

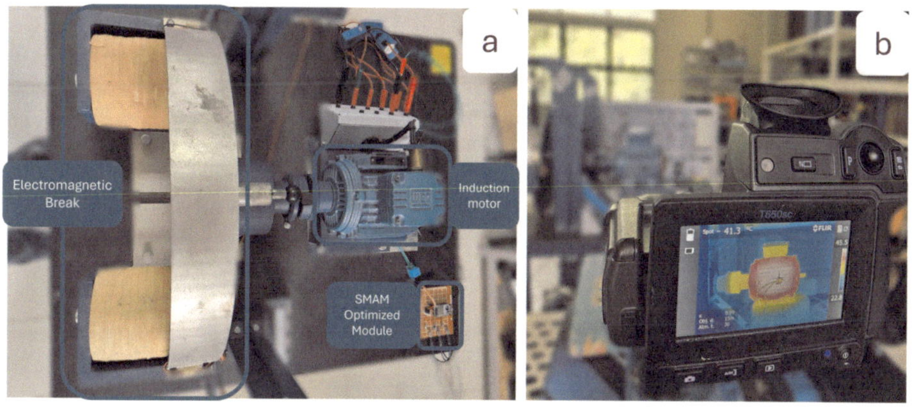

Figure 5.
Arrangement of equipment for testing the embedded system (a) and thermographic camera used in the experiment (b).

2.3 Functional testing methodology

The motor selected for this study is WEG's high-performance Plus model, with a rated current of 4.42 A. This motor operates in conjunction with an electromagnetic break, which consists of coils and a metal disk controlled by an electrical control bench. The function of the brake is to act as a load, simulating real motor operation in industry. The connection between the motor and the brake can be seen in **Figure 5a**.

To test operation, the motor was monitored by the optimized SMAM module and the ET-5062 energy analyzer, used as a reference meter. According to the manufacturer [22], this analyzer complies with the IEC/EN61010-1 standard, Category IV 600 V Overvoltage, and allows the user to choose between 1-second and 60-minute measurements. It also offers real-time visualization of general network parameters such as voltage, current, and power. It is worth noting that the motor is monitored simultaneously with both devices so that measurements can be compared.

The test followed the methodology adopted by Santos et al. [12], sequentially applying four load levels, identified as load 1, 2, 3, and 4. At the initial load (load I), the motor operated at no load, without the Eddy Brake coils energized. At load levels 2, 3, and 4, the motor's phase currents reached values of 3 A, 3.5 A, and 4 A, respectively. **Figure 5** shows the test layout.

To monitor the external temperature of the engine, a FLIR T650sc thermographic 50 camera was used (**Figure 5b**), capable of measuring temperatures from 40 to 650°C. In addition, a Minipa multimeter model ET-2042E was used, with a thermal sensitivity of −40 to 1000°C [23]. The data was stored manually, with the measurements from the reference equipment being compared with the measurements from the optimized module's temperature sensor at each load transition.

2.4 Web application structure

Communication is carried out *via* a web application developed in the Python, HTML, and JavaScript programming languages and stored on the Replit online platform. Replit is a platform that stores web applications and offers resources for testing and deploying applications in various programming languages [24].

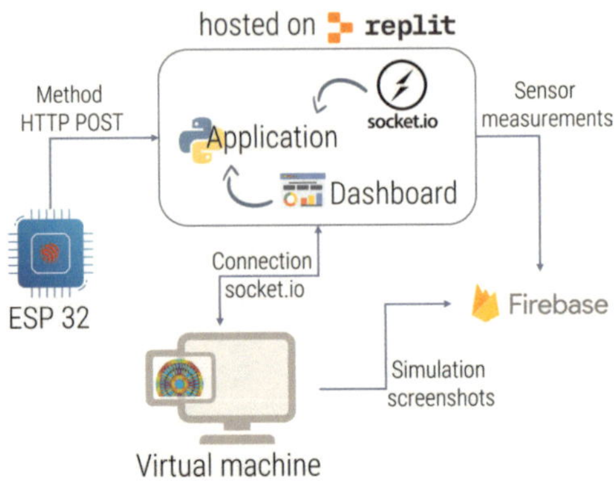

Figure 6.
Structure of the application developed by [12].

The web application receives the data sent by ESP32 in JSON format, using the HTTP POST method. The measurements received are stored in Firebase, a non-relational database from Google that offers various functionalities, including a real-time database [25]. The measurements are stored in Firebase in JSON format.

Socket.io middleware was also implemented in the application. This technology enables real-time communication between the registered client and the server [26]. Thus, as soon as the measurement data is received by the application, the socket connection sends this data to a Virtual Machine (VM).

The VM uses the Windows operating system, as the FEMM software used for computer simulation is native to this system. This VM was configured to detect variations in the motor current and perform a thermomagnetic simulation based on these variations. The result of the simulation is sent to Firebase. The general flow diagram of the communication described is shown in **Figure 6**.

The control panel, developed in HTML and JavaScript, is integrated into the application to display the results of the motor current simulations in real time. Once the simulations are complete, images are sent from the VM to Firebase and the application *via* a socket connection. The JavaScript panel then automatically checks and displays the results on the web page, eliminating the need for the user to manually update the page.

2.5 Thermomagnetic simulation of an induction motor

This research involved performing a multiphysics thermomagnetic simulation utilizing FEMM software. By coupling these simulations, the outcomes from one serve as inputs for another, thereby enhancing the fidelity to actual physical phenomena [27]. The induction motor model features a rotor with 28 slots and a stator with 36 slots, composed of aluminum and copper, respectively.

The simulation process is automated through a Python script leveraging the pyFEMM package, facilitating easy execution of simulations in FEMM. The script defines the motor materials, creates the finite element mesh, and initiates the magnetic simulation. In this stage, Eq. 1 is numerically solved to find the potential of the three magnetic vectors within all generated finite elements.

$$\nabla \times \left(\frac{1}{\mu(\boldsymbol{B})} \nabla \times \boldsymbol{A} \right) = -\sigma \frac{\partial \boldsymbol{A}}{\partial t} - \sigma \nabla V + \boldsymbol{J}_{\text{src}} \tag{3}$$

In this context, $\mu(\boldsymbol{B})$ denotes the magnetic permeability of the medium, and $\boldsymbol{J}_{src}$ represents the current density applied by an external source. The resistive losses in the copper conductors are calculated using Eq. 4.

$$q = \frac{1}{\sigma} J^2 \tag{4}$$

Subsequently, the thermal simulation phase commences, solving Eq. 5 numerically across all finite elements. This equation incorporates the resistive losses from the magnetic simulation as heat sources.

$$\rho c_p \frac{dT}{dt} - \nabla \cdot (k \nabla T) = q \tag{5}$$

Here, ρ_0 is the mass density, c_p signifies the specific heat capacity, k represents the thermal conductivity, and T stands for temperature. The electrical conductivities are updated in accordance with Eq. 6. Where ρ_0 is the electrical resistivity at 0°C, and β_0 is the temperature coefficient of resistivity.

$$\sigma(T) = \frac{1}{\rho_0 (1 + \beta_0) T}, \tag{6}$$

3. Results and discussions

The test carried out to verify the performance of the SMAM's optimized embedded system was satisfactory, since both the ESP32 microcontroller received all the sensor measurements and sent them to Firebase, and the ET-5062 stored all the measurements in its internal memory. In addition, during the test, as described in Section 2.3, the temperature reached by the engine was monitored and it was found that there was no need to apply statistical techniques for calibration, since this sensor showed an average error of around 1%. **Figure** 7 shows the thermogram of the temperature measurement for load 3.

With the testing stage successfully completed, we proceeded to calibration, in which the script described in the previous section came into play, transforming the files generated into a data frame by the Pandas library for studying the data. The first

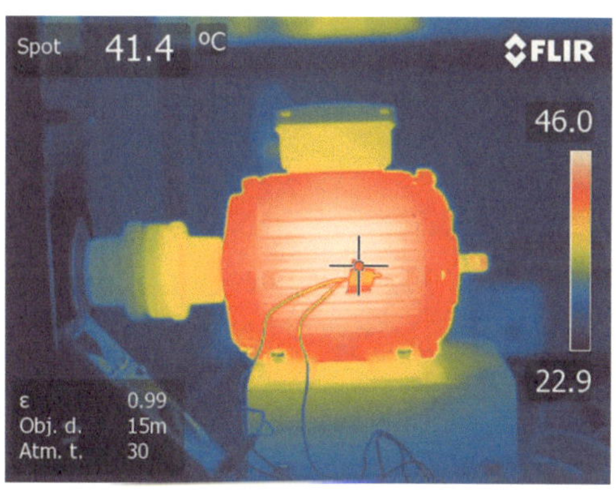

Figure 7.
Thermogram of the motor operating with load 3 in the performance test of the optimized SMAM module.

	Without calibration		With calibration	
Monitored phases	MAE	MAE %	MAE	MAE %
Phase A	0.5992	14.87	0.0681	1.88
Phase B	0.6518	15.48	0.0622	1.79
Phase C	0.5944	14.26	0.0367	1.00

Table 1.
Absolute error of the SMAM current sensors.

analysis made of this was the voltage analysis, in which the mean absolute error of the voltage read by the SMAM and ET-5062 sensor was calculated, the result of which was a mean absolute error (MAE) of 1.5, around 1%. The errors presented by the voltage and temperature sensor follow those described in the work by Santos et al. [12].

The second analysis took place with the current data, in which the mean absolute error was first calculated without calibration to assess the need to apply static techniques. The result of this is shown in **Table 1** and as can be seen the errors presented by each sensor are high compared to the voltage and temperature measurements.

Once the difference between the sensors was noted, the *LinearRegression* functions and the *fit* method were applied to determine the equations of the regression line for each sensor, which resulted in Eqs. 7, 8, and 9.

$$y_i = 0.83x + 0.08 \tag{7}$$

$$y_i = 0.82x + 0.10 \tag{8}$$

$$y_i = 0.85x + 0.04 \tag{9}$$

After this, the calibrated measurements were calculated using the equations found, in which the measurement made by the SMAM's current sensor is the input to the regression model represented by the x variable. After carrying out this process, the average absolute error of the calibrated measurements was measured to indicate the effectiveness of the model and the result obtained is that after using the static techniques the error decreased significantly, since it is below 2% as seen in **Table 1**.

In addition, **Figure 8** shows the graphs for each phase relating the SMAM measurements to the ET-5062 data. As you can see, both the SMAM and the ET-5062 did

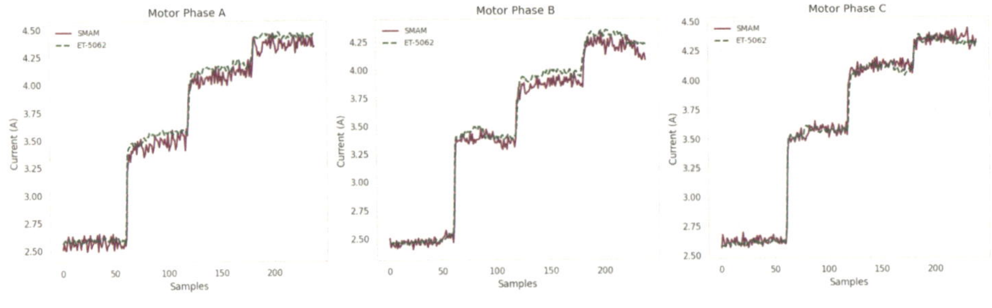

Figure 8.
Graph showing the measurements of the motor's three phases made by the ET-5062 and the SMAM after calibration.

DOI: http://dx.doi.org/10.5772/intechopen.1006914

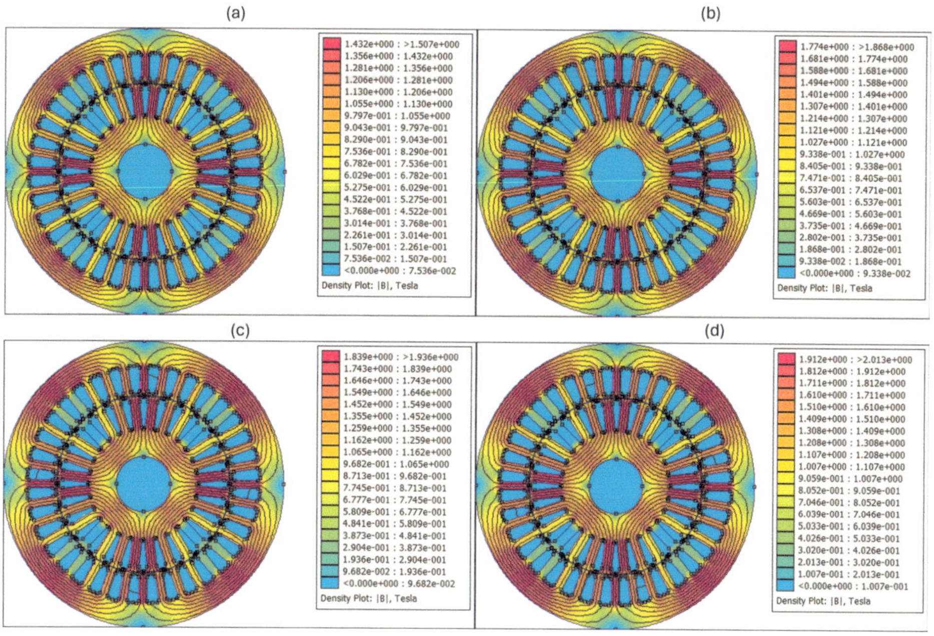

Figure 9.
The magnetic flux of the electric motor when operating at load I (a), load II (b), load III (c), and load IV (d).

not capture the motor's peak current, only the permanent regime in which each step represents the four load levels described in Section 2.3.

As for the magnetic simulation results, it was possible to visualize the magnetic flux present in the stator and rotor, as well as the formation of the four poles inside the motor, using a triangular mesh with 46,426 nodes and 92,294 elements. It was also found that the maximum magnetic induction was 1.5 T, 1.8 T, 1.9 T, and 1.95 T for loads I, II, III, and IV, respectively. As can be seen in **Figure 9**.

In the thermal simulation, the values of the resistive losses in the rotor and stator windings calculated in the magnetic simulation are incorporated into the thermal

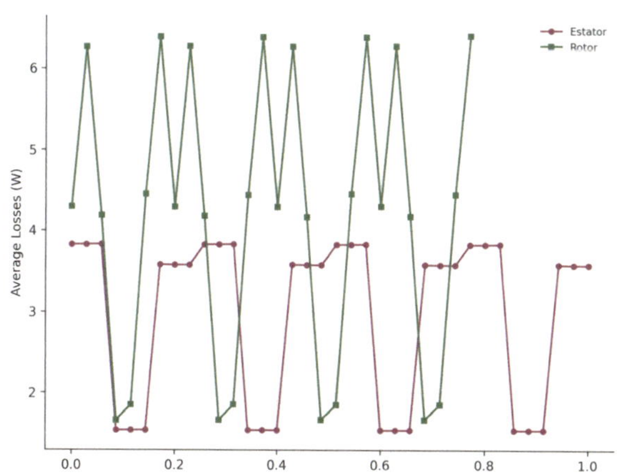

Figure 10.
Resistive losses calculated in the simulation of the motor operating under load IV.

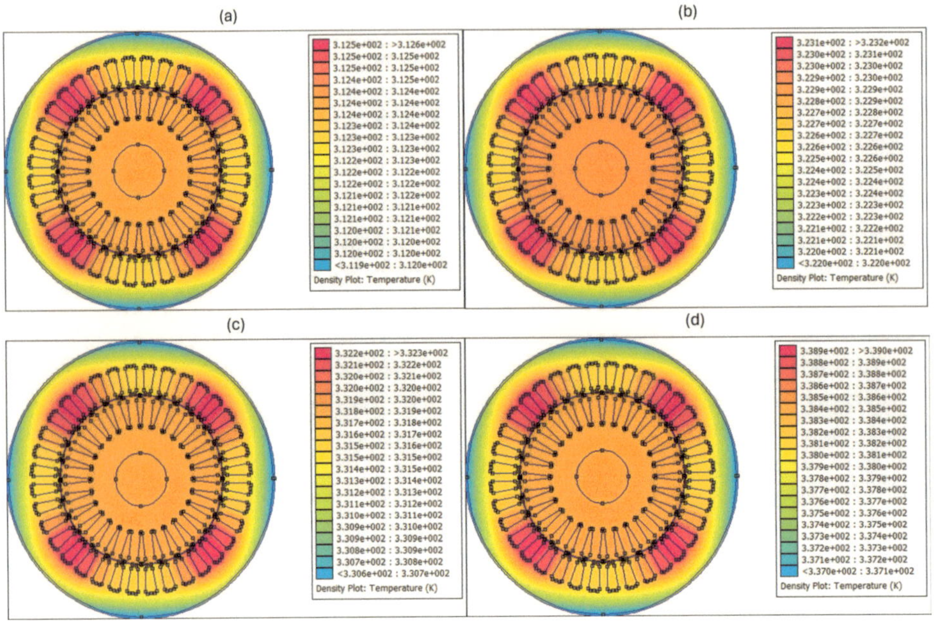

Figure 11.
Temperature in Kelvin of the engine when operating with load I (a), load II (b), load III (c), and load IV (d).

simulation as described above. The graph in **Figure 10** shows the resistive losses (in watts) per slot in the stator and rotor for all loads.

Figure 11 shows the thermal behav of the motor for all loads, and shows the concentration of heat in the center of the poles due to resistive losses and the high concentration of magnetic flux.

In addition, by carrying out the thermal simulation, it is possible to calculate the conductivity of the materials that make up the motor and which are influenced by the temperature which, when it increases, decreases the conductivity of the materials and causes the resistance to increase. **Figure 12** shows the reduction in conductivity in the motor's stator and rotor as the load increases.

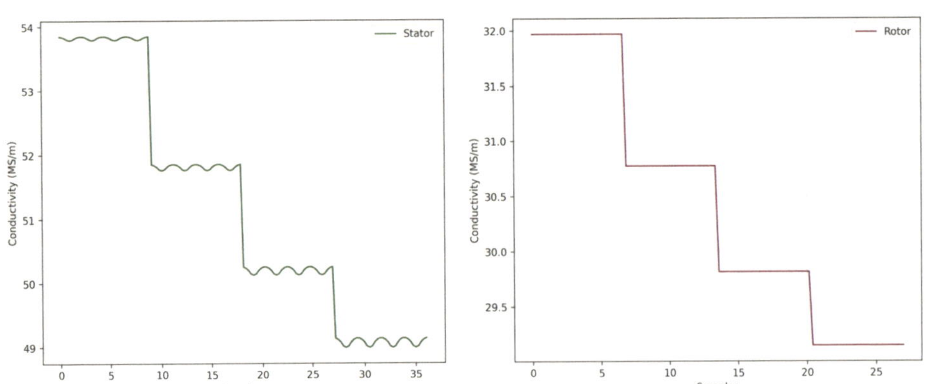

Figure 12.
Electrical conductivity of the stator and rotor.

Finally, as in Ref. [12], the result of each simulation was visualized in the dashboard developed. In this way, it was possible to follow the status of the monitored motor in real time, since the panel is web responsive, defined as a page that adapts to the type of device used according to França [28], it was possible to monitor the thermal and magnetic performance of the motor in operation on both smartphones and computers.

4. Conclusions

This work proposed improvements in the measurements of the sensors present in the Motor Monitoring and Analysis System (SMAM), a Digital Twin (DT) system that can assist in the predictive analysis of faults in three-phase electric motors. The focus was on optimizing the measurements of the current sensors responsible for monitoring the motor's three phases, using mathematical methods to adjust the sensors and validate the results with reference equipment.

The results obtained showed that the proposed improvements were effective, since after calibration, the measurements showed errors of less than 2%. Thus, the performance test of the new SMAM module was satisfactory and provided accurate and reliable measurements. As a result, this work contributes to the advancement of IIoT in industry by showing effective methods for adjusting and calibrating sensors, which is essential for the successful implementation of IIoT and for improving the efficiency and reliability of industrial processes.

Acknowledgements

The authors of this article would like to thank the Financier of Studies and Projects (FINEP) of the Ministry of Science, Technology and Innovation (MCTIC) for its support in carrying out this work, within the scope of the Project entitled "Technological Development and Innovative Extension of Advanced Materials in Energy and Mobility applied to the Amazonian context" (EMOB-AMAZON), carried out through the National Science and Technology Plan of Public Call MCTI/FINEP/AT—Advanced Materials and Strategic Minerals 2020 nº 066/21. The authors would also like to thank the Coordination for the Improvement of Higher Education Personnel (CAPES) and the Center of Excellence in Energy Efficiency in the Amazon (CEAMAZON).

Author details

Jhennifer F. dos Santos[1*†], Yasmim M.D. Lisboa[1†], Elielson F.S. Barbosa[2], Lucas H.B. Santos[1], Allan R.A. Manito[1], Marcelo O. Silva[1] and Wellington S. Fonseca[1]

1 Institute of Technology, Federal University of Pará, Belém, Pará, Brazil

2 University Center of Pará, Belém, Pará, Brazil

*Address all correspondence to: jhennifer.santos@itec.ufpa.br

† These authors contributed equally.

References

[1] Tiwari D, Miscandlon J, Tiwari A, Jewell GW. A Review of Circular Economy Research for Electric Motors and the Role of Industry 4.0 Technologies. Sustainability. 2021;**13**: 9668. DOI: 10.3390/su13179668

[2] Antonino-Daviu J. Electrical monitoring under transient conditions: A new paradigm in electric motors predictive maintenance. Applied Sciences. 2020;**10**(17):6137

[3] Subhashini N, Mouli M, Mugunthan J, Kumar RP, Revanth M, Tejaa M. Automated Induction Motor Monitoring System Using IoT. In: 2023 9th International Conference on Advanced Computing and Communication Systems (ICACCS), Coimbatore, India; 2023. pp. 166-169. DOI: 10.1109/ICACCS57279.2023.10112942

[4] Patel RA, Bhalja B, Alam MA. Condition Monitoring of Three-Phase Induction Motor. In: 2020 IEEE 1st International Conference for Convergence in Engineering (ICCE), Kolkata, India. 2020. pp. 16-20. DOI: 10.1109/ICCE50343.2020.9290540

[5] Noyjeen E, Tanita C, Panthasarn N, Chansri P, Pukkham J. Monitoring Parameters of Three-Phase Induction Motor Using IoT. In: 2021 9th International Electrical Engineering Congress (iEECON), Pattaya, Thailand; 2021. pp. 483-486. DOI: 10.1109/iEECON51072.2021.9440368

[6] Gokilakrishnan G, Dhanamurugan VMA, Bhasha A, Subbiah R. A Review of Applications, Enabling Technologies, Growth Challenges and Solutions for IoT/IIoT. In: 2023 9th International Conference on Advanced Computing and Communication Systems (ICACCS), Coimbatore, India. 2023. pp. 2241-2250. DOI: 10.1109/ICACCS57279.2023.10112825

[7] Singh I, Hodgins S, Gadhrri A, Schmidt R. Iiot machine health monitoring models for education and training. In: Singh I, Gao Z, Massarelli C, editors. IoT Applications Computing. London, UK: IntechOpen; 2021. p. ch. 12. DOI: 10.5772/intechopen.99032

[8] Javaid M, Haleem A, Singh RP, Rab S, Suman R. Upgrading the manufacturing sector via applications of Industrial Internet of Things (IIoT). Sensors International. 2021;**2**:100129. DOI: 10.1016/j.sintl.2021.100129

[9] Sharma S, Monga C. Introduction to Different Simulation Techniques of Digital Twin Development. In: Anand A, Sardana A, Kumar A, Mohapatra SK, Gupta S, editors. Simulation Techniques of Digital Twin in Real-Time Applications: Design Modeling and Implementation. 2024. pp. 1-24

[10] Grieves MW. Digital twins: Past, present, and future. In: The digital twin. Cham: Springer International Publishing; 2023. p. 97-121

[11] Singh RR, et al. Building a digital twin powered intelligent predictive maintenance system for industrial AC machines. Machines. 2023;**11**(8):796

[12] Santos JF, Lisboa YMD, Santos LH, Barbosa E, Manito AR, Fonseca WS, et al. Three-Phase Induction Motor Monitoring System for Predictive Failure Analysis Based on Digital Twin, IIoT and Finite Element Method. In: 2023 15th IEEE International Conference on Industry Applications (INDUSCON), São Bernardo do Campo, Brazil; 2023.

pp. 766-773. DOI: 10.1109/INDUSCON58041.2023.10374867

[13] Espressif Systems. ESP32-WROOM-32D and ESP32-WROOM-32U Datasheet. 2023. Available from: https://encr.pw/22pzH [Accessed: March 25, 2023]

[14] WEG. WCD-GT100 and WCD-GT100-ETH. 2022. Available from: https://static.weg.net/medias/downloadcenter/ha9/ha2/WEG-GATEWAY-GT100-manual-do-usuario-10009992380-pt.pdf [Accessed: March 25, 2023]

[15] YHDC. Split Core Current Transformer Model SCT013–100 Datasheet. 2015. Available from: https://en.yhdc.com/comp/file/download.do?id=941 [Accessed: March 25, 2023]

[16] Qingxian Zeming Langxi Electronic. ZMPT101B Current-Type Voltage Transformer. 2020. Available from: https://5nrorwxhmqqijik.leadongcdn.com/ZMPT101B+specification-aidijBqoKomRilSqqokpjkp.pdf [Accessed: April 23, 2023]

[17] Liang J. Multivariate linear regression method based on SPSS analysis of influencing factors of CPI during epidemic situation. In: 2020 2nd International Conference on Economic Management and Model Engineering (ICEMME), Chongqing, China; 2020. pp. 294-297. DOI: 10.1109/ICEMME51517.2020.00062

[18] Pengfei L, Shun T, Liting Y, Yeniu Q, Jian S. An assessment method of power system harmonic impedance based on mixed total least squares. In: 2016 IEEE PES Asia-Pacific Power and Energy Engineering Conference (APPEEC), Xi'an; 2016. pp. 1694-1698. DOI: 10.1109/APPEEC.2016.7779839

[19] Chen DY. Análise de Dados Com Python e Pandas, 1st ed. São Paulo, Brazil: NOVATEC; 2018; ISBN 978-85-7522-699-5

[20] Nelli F. Data Visualization with matplotlib and Seaborn. In: Python Data Analytics: With Pandas, NumPy, and Matplotlib. Berkeley, CA: Apress; 2023. pp. 183-257

[21] Haposan D, Hidayat MN, Santosa A, Ibrahim MA, Majiid RN. Estimated Twitch Streamer Revenue Using Linear Regression Algorithm. In: 2023 5th International Conference on Cybernetics and Intelligent System (ICORIS). Pangkalpinang, Indonesia; 2023. pp. 1-5. DOI: 10.1109/ICORIS60118.2023.10352194

[22] Minipa Electric. Analisador de Energia Modelo ET-5062. 2020. Available from: https://www.minipaelectric.com.br/images/propostatecnica/ET50621313BR.pdf [Accessed: April 10, 2024]

[23] Minipa Electric. Multimetro Digital Modelo: ET-2042E. 2016. Available from: https://www.minipa.com.br/images/Manual/ET-2042E-1101-BR.pdf [Accessed: April 10, 2024]

[24] Replit. Replit. 2023. Available from: https://replit.com/ [Accessed: April 29, 2023]

[25] Firebase. Firebase Realtime Database. 2023. Available from: https://acesse.one/jckNW [Accessed: April 29, 2023]

[26] Socket.IO. What Socket.IO Is. 2023. Available from: https://socket.io/docs/v4/ [Accessed: April 29, 2023]

[27] Wang Y, Li S, Meng W. Strong coupling analysis of fluid–solid for magnetorheological fluid braking

system. Journal of Intelligent Material Systems and Structures. 2018;**29**(8): 1586-1599

[28] França SDS. Web design responsivo: Caminhos para um site adaptÁvel. Interfaces Científicas—Exatas e Tecnológicas. 2015;**1**(2):75-84. Available from: https://periodicos.set.edu.br/exatas/article/view/2220

Chapter 2

Open Circuit Fault Diagnosis in Induction Motor Driver Inverter

Francisco Javier Villalobos-Pina, Josué A. Reyes-Malanche, Eduardo Cabal-Yepez and Efrain Ramirez-Velasco

Abstract

The induction machines are the power horses in the industry due to the practically null maintenance, this kind of machines are use in a widely group of industrial applications, and with the advance of power electronics these machines replace another kind, like direct current (dc) motors attributable to the evolution of control algorithms and the digital platforms. In this context, a methodology is proposed to detect and isolate faults, focusing on the inverter stage of Induction Machine motor drives, with a specific emphasis on Insulated Gate Bipolar Transistor (IGBT) faults, using phasor analysis and fuzzy logic. This methodology has demonstrated effective performance in detecting and isolating different types of electrical faults, such as stator inter-turn short-circuit. In this case, damaged switching IGBTs were identified using low computational resources. This research was motivated because some complex techniques like music (multiple signal classifier), dynamic observers, techniques based on mathematical models, statistics techniques, optimization techniques, AI techniques like deep learning, vector support machines, genetic algorithms, and so on, require a great quantity of data and or computer processing. Thanks to this scheme, it is possible to implement a low cost computational platform based on a TI DSP TMS320F28335 processor for a real-time fault diagnosis in Induction Machine inverter.

Keywords: induction machines, instantaneous space phasor (ISP) module, Fourier transform, wavelet transform, current phasor, fuzzy logic

1. Introduction

Induction motors are a type of electric motor known for their trustworthiness and applications in different processes. However, like any other electromechanical system, they are subject to different types of faults throughout their useful life. In this regard, various methodologies have been proposed to detect faults in different components, such as the motor driver, which is a critical component in the functioning of these motors.

There are numerous works in this field, and research on the permanent magnet (PM) brushless DC motor (BLDC) drives is analyzed under single and multiple fault conditions. The BLDC motor is powered by a three-phase and two-level inverter with six switches using IGBT technology. The three-phase source currents are used as fault signatures. An IA algorithm can detect and identify the open-circuit faulty switch. The proposed algorithm's effectiveness and simulation results are shown in [1].

On the other hand, an algorithm is proposed for BLDC motor drives to maintain control performance under switch open damage. Using the measured current phase, this algorithm acquires the stator currents to detect open-circuit faulty conditions. The algorithm reaches a rapid control performance and reconfigures the system topology [2].

Another research is based on voltage switching patterns of IBGT, and the first proposed scheme is based on phase voltage, and the second is based on the line voltage of the inverter. This methodology detects single and multiple open circuit switch fault inverter scenarios. A computational analysis of cost-effectiveness- between both methods is performed, and reach ns is performed in the best case. The diagnostic time is better for the first method than the second for one and a half of the fundamental voltage period. Experimental results are presented in Ref. [3].

Conversely, an open circuit switch fault scheme in pulse width modulated inverters is proposed. The scheme uses normalized variables and acquires only two inverter currents for a closed-loop control. The algorithm uses the line current shapes and the α and β reference frame under healthy and faulty operation conditions. The diagnosis scheme first uses 27 acquired drive current patterns to detect the faulty inverter legs. Only six patterns were used to detect a single fault; the rest were for simultaneous faults [4].

Lastly, a methodology to choose suitable is the use of entropy variation and parameter values. This proposed algorithm studies a BLDC motor powered by a three-phase inverter, and single and simultaneous faults are detected in a switch open circuit. The proposed scheme uses atypical characteristics of phase currents, and a mathematical model is used to select the appropriate entropy functions to detect accurately the inverter open-circuit faults [5].

This work proposes detecting and isolating open-circuit faults on the switching IGBTs in Induction Motor (IM) Drive through digital signal processing techniques such as Park instantaneous space phasor (ISP) modulus, which mapped the IM stator currents to α and β reference frame. Spectral analysis uses the fast Fourier transform (FFT) tool, multiresolution analysis (MRA) through Haar level one wavelet transform, and phasor and fuzzy logic analysis of stator line currents.

2. System description

An experimental test bench, designed for studying electrical faults in power switches, is employed for analyzing electric and electronic issues on the power converter as well as on the IM (Induction Machine) machine, as shown in **Figure 1**.

The experimental system is composed of a high-efficiency *Siemens* 220-volt, three-phase induction motor, with a diameter of 10 inches, of 2.25 kW, 2-pole-pairs, and 1800 rpm maximum speed, a generator with a resistive load to produce torque, a

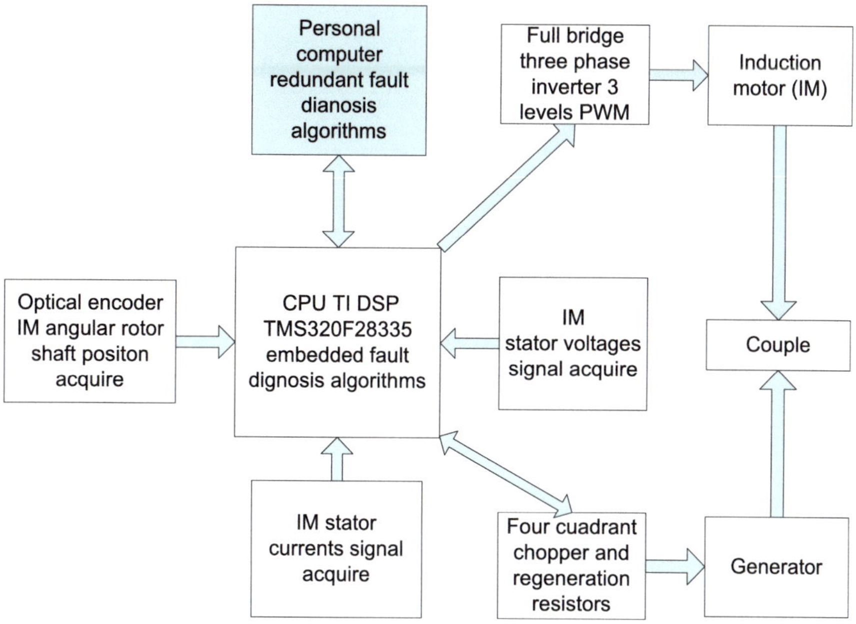

Figure 1.
IM machine test bench schematic diagram.

three-level, full-bridge, electronic power system, a data acquisition board based on close-loop hall-effect sensors, from FW-Bell Inc., the *CSNA*11, to acquire the stator current signals from the IM, and a 32-bit, a float point unit DSP central processing unit (CPU) based on the TMS320F28335 digital signal controller from Texas Instrument. A software program developed on Delphi-XE7 Embarcadero is used to control and acquire information from the electronic system through a personal computer utilizing a high-speed serial port.

3. Phasor analysis for open-circuit fault detection and isolation

Analyzing an open-circuit fault on one or two power brakers from a single branch of a three-phase inverter feeding a brushless motor, some distinctive behaviors can be observed on the motor power supply currents. Specifically, when an open-circuit fault occurs on the branch corresponding to the $\mathbf{I}_{ca}$ current, the $\mathbf{I}_{ab}$ and $\mathbf{I}_{bc}$ trajectories show a characteristic conduct.

The analytical expressions in Eq. (1) represent the behavior of the power supply currents when an open-circuit fault occurs by taking into account the phasorial representation of the three-phase power-inverter currents under normal operation, without any faulty condition, as depicted in **Figure 2**:

$$\begin{aligned}\mathbf{I}_{ca} &= I_m\angle 0^\circ \\ \mathbf{I}_{ab} &= I_m\angle -120^\circ \\ \mathbf{I}_{bc} &= I_m\angle 120^\circ\end{aligned} \quad (1)$$

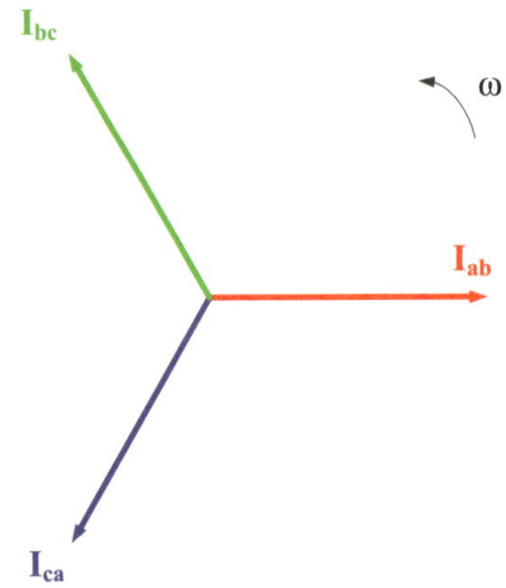

Figure 2.
Current sequence healthy case.

Taking in mind that in a balanced, three-phase system, the condition described in Eq. (2) is always satisfied

$$\mathbf{I}_{ca} + \mathbf{I}_{ab} + \mathbf{I}_{bc} = 0 \tag{2}$$

A specific current phasor can be represented with reference to a coordinate system built up by the phasors of the remaining two currents. In other words, a particular current phasor can be represented through a linear combination of the other two current phasors, as described in Eq. (3):

$$\mathbf{I}_{ab} \quad = \quad \alpha_1 \mathbf{I}_{ca} + \beta_1 \mathbf{I}_{bc} \tag{3}$$

Figure 3 shows $\mathbf{I}_{ab}$ coordinates regarding to the $\mathbf{I}_{ca}$ and $\mathbf{I}_{bc}$ reference system, where $\alpha_1 = -1$ and $\beta_1 = -1$.

When an open-circuit fault takes place on the switch corresponding to the $\mathbf{I}_{ca}$ current, which works as a basis for the reference coordinate system, $\mathbf{I}_{ab}$ moves forward situating at counterphase to $\mathbf{I}_{bc}$, as depicted in **Figure 4**. Therefore, it is possible to identify and isolate open-circuit faults by identifying the power supply currents located at counterphase.

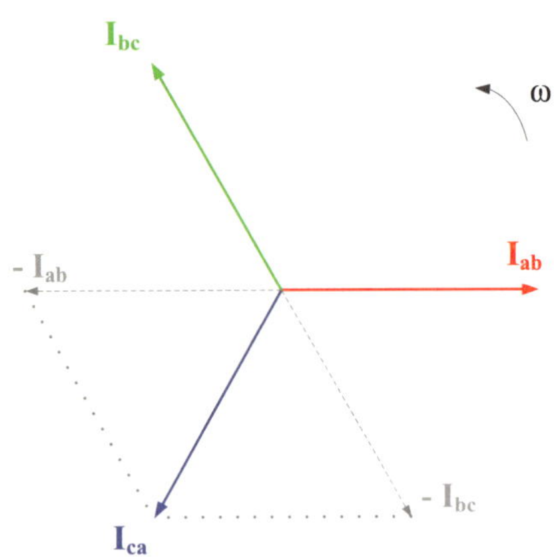

Figure 3.
Current components of $\mathbf{I}_{ca}$ in terms of $\mathbf{I}_{ab}$ and $\mathbf{I}_{bc}$.

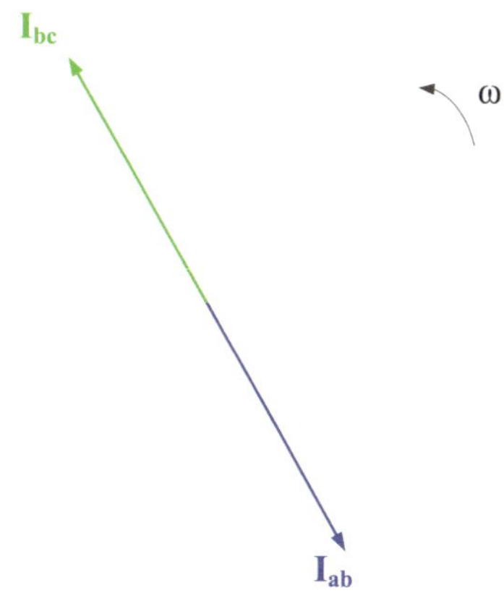

Figure 4.
Currents I_{ab} at counterphase with I_{bc}.

In this way is possible to detect and isolate open-circuit faults depending on which currents are in counterphase.

4. Switch faults in power converters

The power electronic system used for controlling electrical machines is known as driver or inverter because it transforms a signal from DC into AC. This device utilizes electronic switches (transistors, thyristors, etc.) because of their tolerance to rough operational conditions; however, the harsh commutation process will wear the devices and produce failures, which will provoke power quality disturbances that degrade the power system itself, its efficiency, and affect the electrical machine performance. In this regard, the two most common failures are:

- Short-circuit faults.
- Open-circuit faults.

A short-circuit fault occurs when any of the switches, building up the power bridge or inverter, allows the electric current to go through without limitation. On the other hand, an open-circuit fault appears when any of the switches in the power bridge interrupts the electric current flow through it, affecting the IM power supply.

Different fault-diagnosis techniques on power switches are assessed by performing 30 different trials on the healthy and faulty conditions of the power bridge feeding the IM motor. The electric power current supply, feeding a 2-pole-pairs IM motor with a maximum rotating speed of 1800 RPM, is acquired at 3.840 kHz, obtaining 128 discrete samples.

4.1 Experimental open-circuit fault diagnosis

Distinct fault scenarios are considered to evaluate experimentally obtained results. A first faulty scheme contemplates an open-circuit fault on the F_A-high switch of phase A, which consists in taking away the electrical gate signal of an IGBT transistor throughout the DSP firmware program. A second scenario considers coincident faulty conditions occurring on F_A-high and F_A-low switches of phase A. The results from both faulty events are contrasted against those from a healthy case, as shown in **Figure 5**. **Figure 6** depicts the structure of the full-bridge, three-phase inverter utilized for controlling the IM of a lightweight electric vehicle.

A common faulty condition on the power electronic system of a motor drive is the open-circuit condition. This kind of defect allows the IM to continue to work under certain rotating speed conditions, provoking the deterioration of the motor and the electronic power system. **Figure 7** shows the experimentally acquired stator current for the first faulty scenario, where the F_A-high switch of phase A is detached, and the stator line current $\mathbf{I}_{ca}$ demonstrates it. In **Figure 7**, it can also be observed that when $\mathbf{I}_{ca}$ equals zero, $\mathbf{I}_{ab}$ and $\mathbf{I}_{bc}$ become in counterphase.

The second faulty condition consists of cutting off the current through switches F_A-high and F_A-low simultaneously. The experimentally acquired stator currents

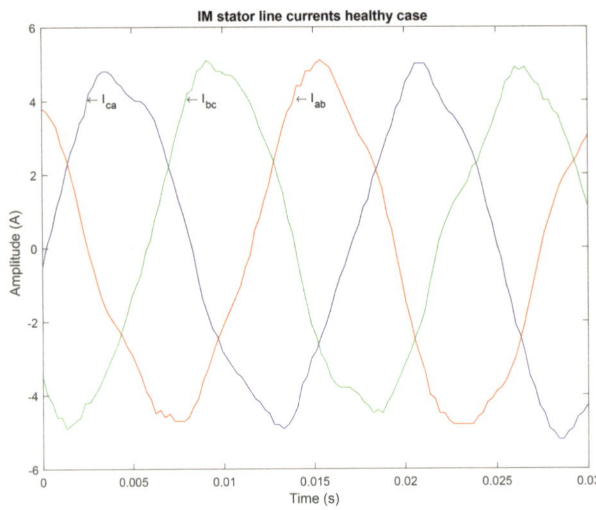

Figure 5.
Stator currents healthy case.

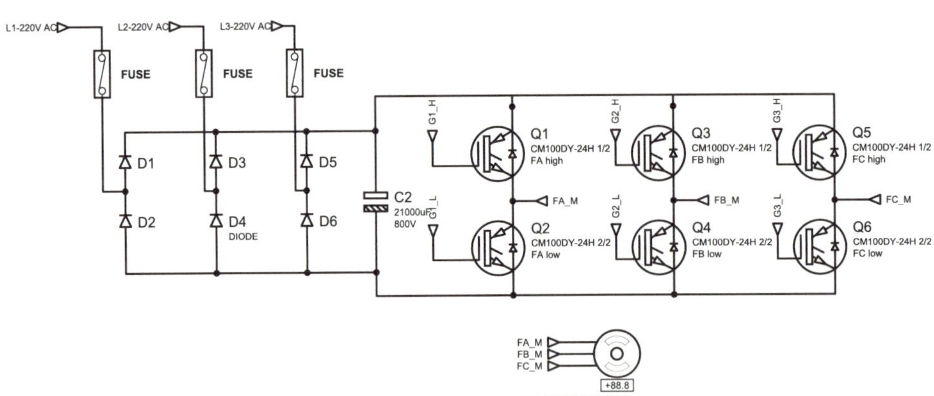

Figure 6.
IM Power Converter Schematic Diagram.

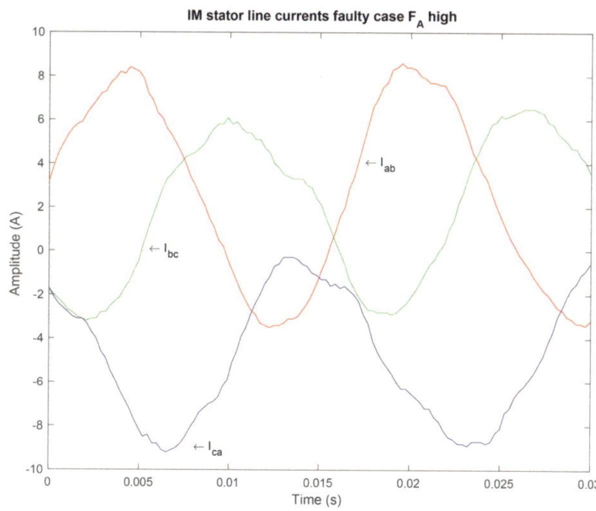

Figure 7.
Stator currents faulty case F_A high.

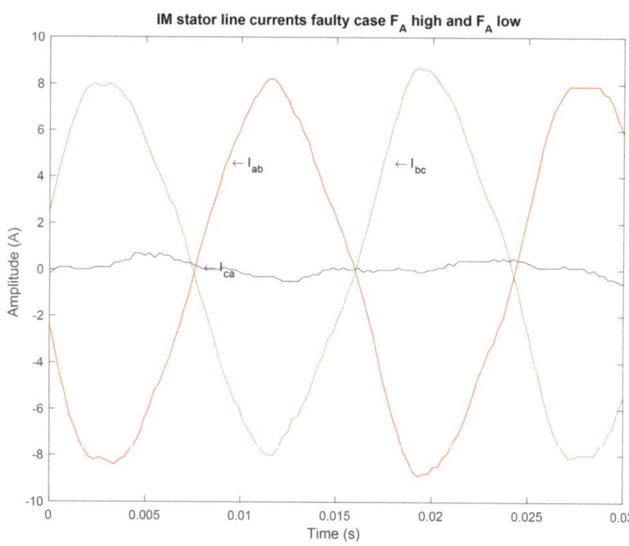

Figure 8.
Stator currents faulty case F_A high and F_A low.

under this operational condition are shown in **Figure 8**, where it can be observed that, as described in Section 3, phasor analysis for open-circuit fault detection and isolation, the current $\mathbf{I}_{ca}$ becomes zero and currents $\mathbf{I}_{ab}$ and $\mathbf{I}_{bc}$ have opposite phase.

5. Park's instantaneous space phasor analysis

The Park's Instantaneous Space Phasor (ISP) transformation is given by Eq. (4), where i_{ab}, i_{bc}, and i_{ca} are the instantaneous three-phase stator currents of the IM. For the fixed reference frame or Concordia transformation, the instantaneous current values in the $\alpha - \beta$ framework are given by i_α, i_β, and i_0 [6].

$$\begin{bmatrix} i_\alpha \\ i_\beta \\ i_o \end{bmatrix} = \frac{1}{\sqrt{2}} \begin{bmatrix} 1 & \cos\left(-\frac{2\pi}{3}\right) & \cos\left(\frac{2\pi}{3}\right) \\ 0 & -sin\left(-\frac{2\pi}{3}\right) & -sin\left(\frac{2\pi}{3}\right) \\ \frac{1}{\sqrt{2}} & \frac{1}{\sqrt{2}} & \frac{1}{\sqrt{2}} \end{bmatrix} \begin{bmatrix} i_{ab} \\ i_{bc} \\ i_{ca} \end{bmatrix} \tag{4}$$

The magnitude of the instantaneous-space-vector modulus can be obtained by

$$\left\|i_{\alpha\beta}\right\|^2 = i_\alpha^2 + i_\beta^2 \tag{5}$$

Figure 9 depicts the ISP modulus, experimentally obtained through the Concordia transformation, from the stator currents of the healthy IM. **Figure 10** shows the ISP modulus from the stator currents collected for the case when an open-circuit fault occurs at the F_A-high switch of phase A. **Figure 11** illustrates the ISP modulus for the faulty case where an open-circuit condition occurs on the F_A-high and F_A-low switches of phase A, concurrently.

5.1 Frequency analysis

The open-circuit fault diagnosis on power inverters feeding IM is performed through the motor current signature analysis (MCSA). A common technique for this purpose examines the instantaneous space vector (ISV), defined in Eq. (5), in the frequency domain. Taking into consideration that the current negative-sequence components are directly associated with the asymmetry provoked by the faulty condition on the power inverter, the presence of this defect can be noticed by analysis of the ISV frequency spectrum [7].

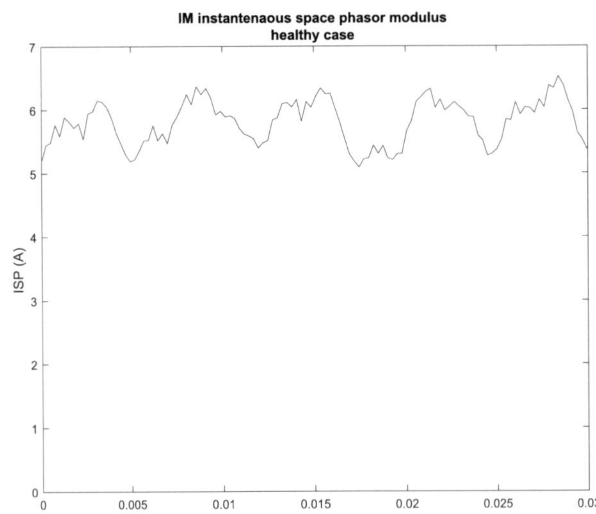

Figure 9.
ISP modulus healthy case.

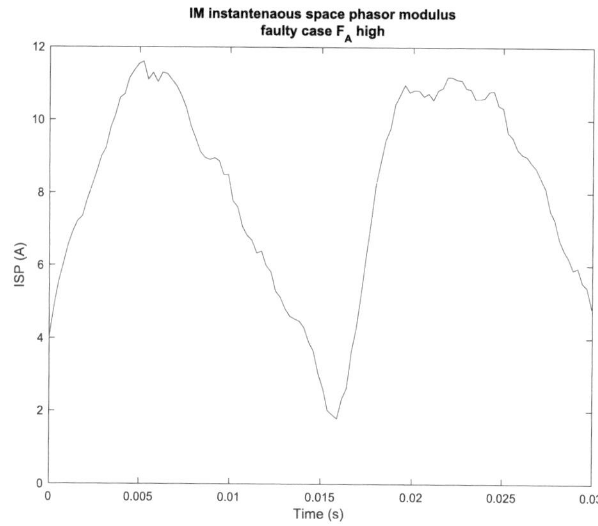

Figure 10.
ISP modulus faulty case F_A high.

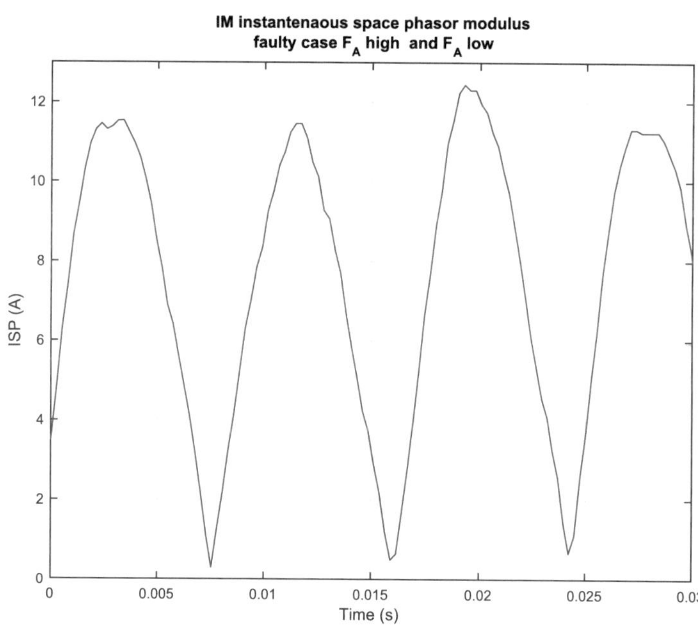

Figure 11.
ISP modulus faulty case F_A high and F_A low.

The open-circuit failure on the power inverter can be located at the frequency F_{fault}, defined by Eq. (6), which will be found at twice inter-multiples $n = 1,2,3, \dots, \infty$ of the line frequency F_L, in the ISV frequency spectrum.

$$F_{fault} = 2nF_L \tag{6}$$

The magnitude of the characteristic fault frequency component F_{fault} decreases as the integer multiple increases; therefore, the fault frequency component at $n = 1$ is considered for useful purposes since it possesses the higher magnitude. Hence, Eq. (6) can be rewritten as Eq. (7).

$$F_{fault} = 2F_L \tag{7}$$

Figure 12 shows the experimentally obtained frequency spectrum for the healthy case, where the main frequency component at $F_L = 60$ Hz, for a rotating speed of 1800 RPM, can be observed. **Figure 13** depicts the frequency spectrum for the open-circuit faulty condition at the F_A-high switch of phase A. **Figure 14** shows the frequency spectrum for the open-circuit condition occurring simultaneously at the F_A-high and F_A-low switches of phase A.

An alternative instrument to carry out fault diagnosis in power inverters is to assess the phasor representations of line currents and phase currents [8]. **Figure 15** depicts the frequency spectrum of this phasor representation when an open-circuit fault occurs on the F_A-high switch of phase A. **Figure 16** displays the frequency spectrum for simultaneous open-circuit faulty conditions on the F_A-high and F_A-low switches. The phasor representation of line and phase currents allows for the diagnosis and discrimination among distinct faulty conditions, different from the ISV, which allows for the identification of the faulty condition. Furthermore, the phasor representation of line and phase currents offers a higher sensitivity during fault diagnosis and classification.

In the **Figures 15** and **16** it can be observed that frequency spectra magnitude changes in the fundamental harmonic $F_{fun} = 60$ Hz that is different of the ISV analysis in which the component spectra appears at the double of the line frequency $2F_{fun} = 120$ Hz.

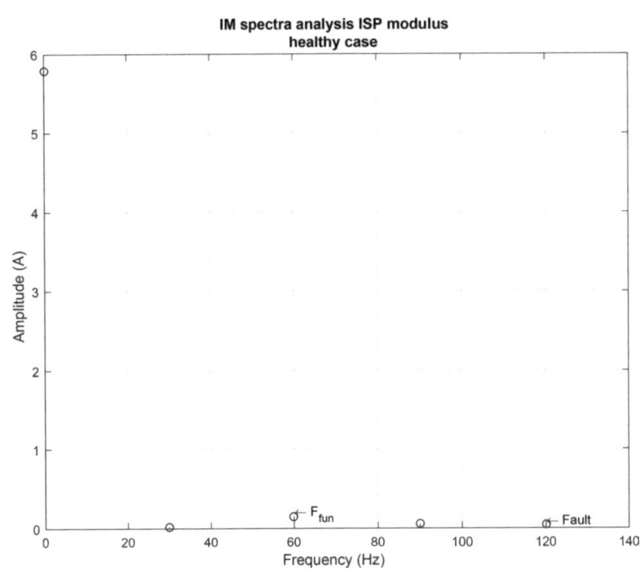

Figure 12.
ISP magnitude spectra healthy case.

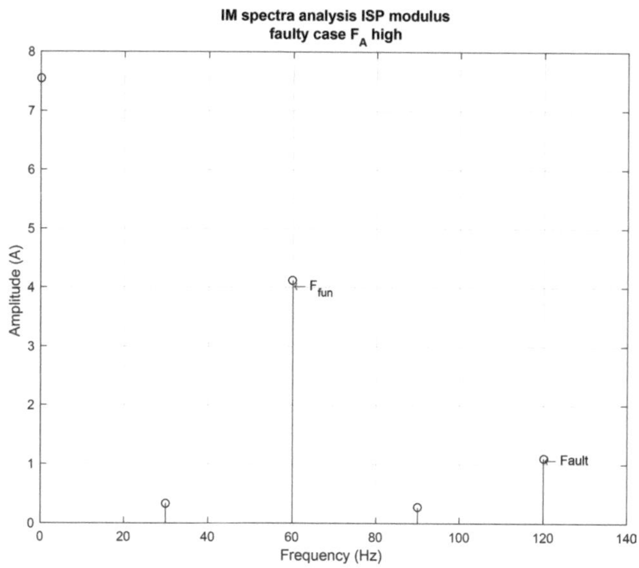

Figure 13.
ISP magnitude spectra faulty case F_A high.

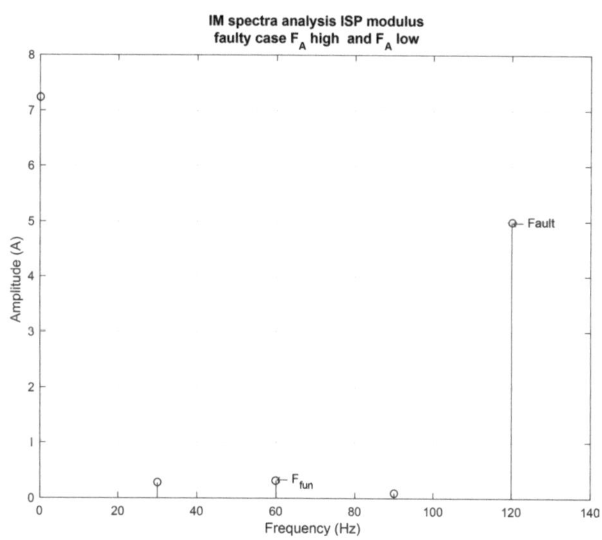

Figure 14.
ISP magnitude spectra faulty case F_A high and F_A low.

5.2 Wavelet analysis

The discrete wavelet transform is an efficient and powerful tool that allows the breaking down of the fault signature contained in the ISV into distinct scales or resolution levels [9]. The wavelet-based multiresolution analysis (MRA) can be used as a redundant examination tool, besides the ISV and the phasor representations of line currents and phase currents, for determining the wavelet coefficients' root

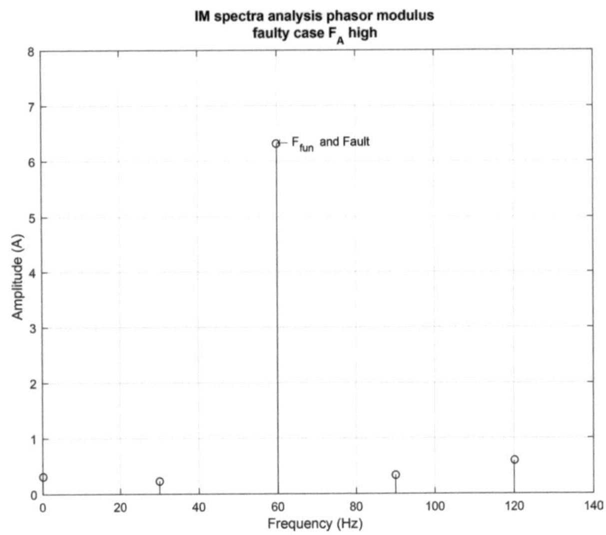

Figure 15.
Phasor magnitude spectra faulty case F_A high.

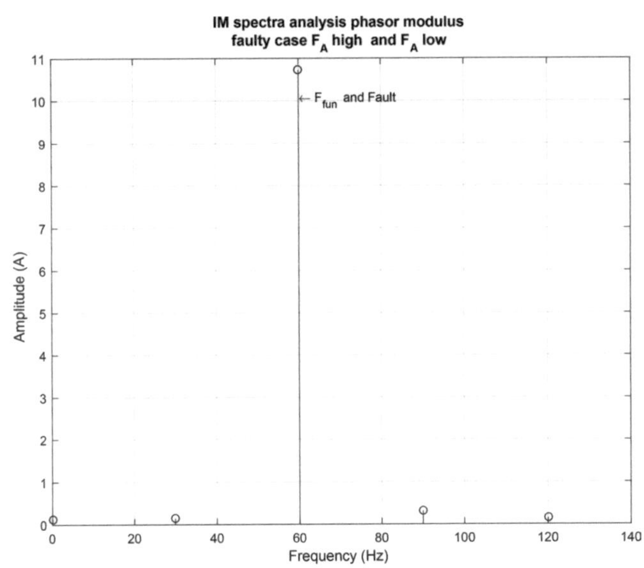

Figure 16.
Phasor magnitude spectra faulty case F_A high and F_A low.

mean square (RMS) value on each frequency band. **Figure 17** depicts the multilevel decomposition.

5.3 Haar mother wavelet function

Level one Haar mother wavelet is the simplest function. It relates transformation in its discrete form, which decomposes one signal into two constituent elements, the trend a_m and fluctuation d_m, with the same length.

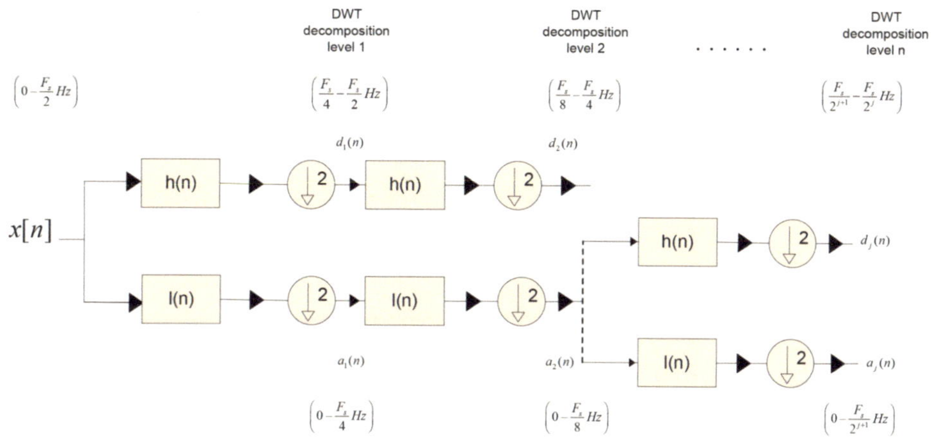

Figure 17.
Wavelet multiresolution analysis scheme.

The approximation coefficients for $m = 1,2,3, \dots, N/2$, where N is the original signal length, are computed by:

$$a_m = \frac{f_{2m-1} + f_{2m}}{\sqrt{2}} \tag{8}$$

The corresponding detail coefficients are given by:

$$d_m = \frac{f_{2m-1} - f_{2m}}{\sqrt{2}} \tag{9}$$

The level 1 Haar transform is computed through:

$$f \overset{H_1}{\rightarrow} (a^1 | d^1) \tag{10}$$

The level 1 Haar mother wavelet is used for obtaining multiple decomposition levels to perform the MRA as follows:

$$\begin{aligned}
Wh1_1^1 &= \left(\frac{1}{\sqrt{2}}, -\frac{1}{\sqrt{2}}, 0, 0, \dots, 0\right) \\
Wh1_2^1 &= \left(0, 0, \frac{1}{\sqrt{2}}, -\frac{1}{\sqrt{2}}, 0, 0, \dots, 0\right) \\
Wh1_2^1 &= \left(0,0,0,0, \frac{1}{\sqrt{2}}, -\frac{1}{\sqrt{2}}, 0, 0, \dots, 0\right) \\
&\vdots \\
Wh1_{N/2}^1 &= \left(0, 0, \dots, 0, \frac{1}{\sqrt{2}}, -\frac{1}{\sqrt{2}}\right)
\end{aligned} \tag{11}$$

The scale functions for the level 1 Haar mother wavelet with multiple decomposition levels are given by:

$$\begin{aligned} Vh1_1^1 &= \left(\frac{1}{\sqrt{2}}, \frac{1}{\sqrt{2}}, 0, 0, \dots, 0\right) \\ Vh1_2^1 &= \left(0, 0, \frac{1}{\sqrt{2}}, \frac{1}{\sqrt{2}}, 0, 0, \dots, 0\right) \\ Vh1_2^1 &= \left(0,0,0,0, \frac{1}{\sqrt{2}}, \frac{1}{\sqrt{2}}, 0, 0, \dots, 0\right) \\ &\vdots \\ Vh1_{N/2}^1 &= \left(0, 0, \dots, 0, \frac{1}{\sqrt{2}}, \frac{1}{\sqrt{2}}\right) \end{aligned} \tag{12}$$

5.4 Experimental multiresolution analysis

The open-circuit faulty condition on the F_A-high switch of phase A in the power converter is inspected meticulously utilizing the MRA of the ISV through the discrete wavelet transform employing the level 1 Haar wavelet mother function. **Table 1** shows a comparative analysis of the effective values (RMS) from the wavelet coefficients corresponding to the distinct ISV frequency bands. This case of study utilizes a MI at 1800 RPM rotating speed. The frequency of the power-supply line is 60 Hz and the fault characteristic frequency is located at $f_{fault} = 120$ Hz; hence, a meaningful change can be observed on the coefficient RMS values in the frequency band from 60 Hz to 120 Hz, which increases from 0.9173 for the healthy case to 10.7920 for the single faulty condition on the F_A-high switch. On the other hand, it is worth noting that the frequency band from 120 Hz to 240 Hz shows some modifications, too, such as increasing the coefficient RMS values from 0.6924 for the healthy case to 3.9965 for the same faulty condition. These two frequency bands are affected because the fault characteristic frequency is located at the boundary between both of them at $f_{fault} = 120Hz$. For the case of simultaneous open-circuit faults on the F_A-high and F_A-low switches, the frequency band between 60 Hz and 120 Hz exhibits a significant alteration in the coefficient RMS values, going from 0.9173 for the healthy case to 9.9142 for the faulty condition. For this case of study, the frequency band from 120 Hz to 240 Hz shows some important modifications too, showing an RMS value of 0.6924 for the healthy case, and an RMS value of 4.9419 for the concurrent faulty conditions.

Level	Frequency band (Hz)	Healthy (RMS)	Faulty F_A high (RMS)	Faulty F_A high and F_A low (RMS)
1	960.0–1920.0	0.1267	0.3038	0.5784
2	480.0–960.0	0.2335	0.7593	1.4634
3	240.0–480.0	0.3848	1.8701	3.9965
4	120.0–240.0	0.6924	3.9965	4.9419
5	60.0–120.0	0.9173	10.7920	9.9142
6	30.0–60.0	0.9836	15.5401	1.8034
7	15.0–30.0	0.2974	1.3355	0.8308

Table 1.
ISV Haar level 1 wavelet MRA coefficients RMS values.

Level	Frequency band (Hz)	Healthy (RMS)	Faulty F_A high (RMS)	Faulty F_A high and F_A low (RMS)
1	960.0–1920.0	0.2914	0.3211	0.5662
2	480.0–960.0	0.8173	0.8826	1.5414
3	240.0–480.0	2.2075	2.4484	4.3738
4	120.0–240.0	5.7460	6.6249	10.3422
5	60.0–120.0	8.5042	8.2157	12.1934
6	30.0–60.0	24.7219	30.3983	52.6997
7	15.0–30.0	0.7896	0.8455	0.8308

Table 2.
Phasor Haar level 1 wavelet MRA coefficients RMS values.

The phasor representation of line currents and phase currents is assessed too through the MRA, utilizing the discrete wavelet transform, for diagnosing and identifying open-circuit faults in a power inverter feeding a MI. **Table 2** shows the experimentally obtained results, where an evident difference between the RMS values of the wavelet coefficients is noticed at the frequency band from 30 Hz to 60 Hz, going from 24.7219 for the healthy case to 30.3983 for the single open-circuit condition on the F_A-high switch. Similarly, for the coincident open-circuit conditions on the F_A-high and F_A-low switches of phase A in the power inverter, an important adjustment on the RMS values of the wavelet coefficients for the frequency band from 30 Hz to 60 Hz can be observed, going from 24.7219 for the healthy case to 52.6997 for the simultaneous open-circuit faults. This significant difference between the RMS values of the wavelet coefficients for the healthy and faulty conditions points out that the examination utilizing the phasor representation of the line and phase currents offers great sensitivity in detecting and classifying open-circuit faults.

6. Fault classification and diagnosis with fuzzy logic

Fuzzy logic can recognize the line-current asymmetries, using the RMS and average (μ) values as data input to the Mamdani algorithm depicted in **Figure 18**, in which it is possible to detect, isolate, and classify this kind of fault.

In **Table 3**, RMS and average values of the Line currents do not show significant changes among them. **Figure 19** shows the obtained results from introducing these values to the fuzzy-logic system, which detects the condition as healthy by providing outcomes of 0.164 in the three currents, using the centroid method are computed. Values from 0 to 1 conform the adopted discourse universe in the proposed fuzzy system as outputs, and it is possible to evaluate healthy IMD (induction motor drive) conditions. The healthy threshold is established from 0 to 0.33 and the faulty thresholds go from 0.33 to 0.66 and 0.66 to 1 for 1 or 2 switches in open-circuit state, respectively.

When a switch sets on an open-circuit state, a semi-cycle of the current signal, corresponding to the faulty condition, is lost, as depicted in **Figure 7**. Therefore, **Table 4** shows the RMS and average values of the line currents $\mathbf{I}_{ca}$, $\mathbf{I}_{ab}$ and $\mathbf{I}_{ab}$, locating at counterphase from each other at the time when $\mathbf{I}_{ca}$ is equal to zero. In this way, the

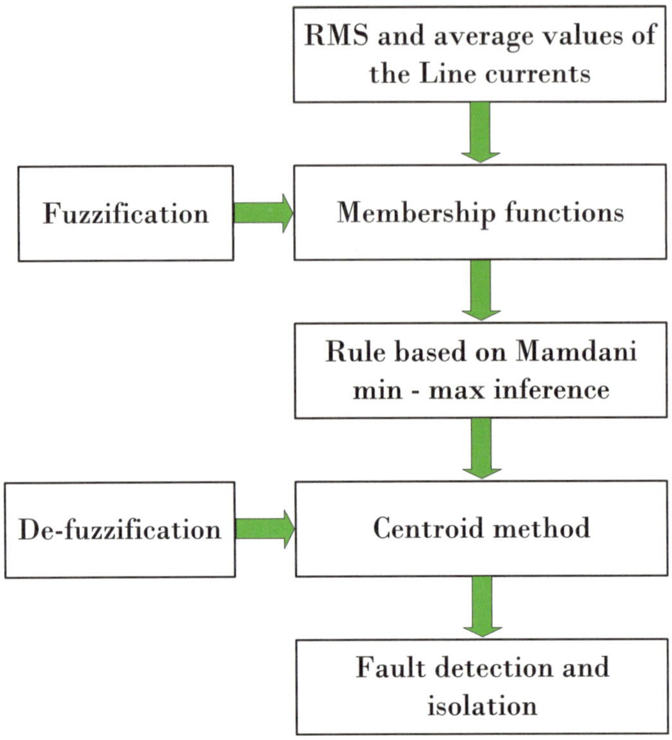

Figure 18.
Fuzzy Logic Algorithm based on Mamdani model.

Trial	$I_{ab}(\mu)$	$I_{ca}(\mu)$	$I_{bc}(\mu)$	$I_{ab}(RMS)$	$I_{ca}(RMS)$	$I_{bc}(RMS)$
1	−0.0836	0.0695	−0.1359	3.3532	3.3263	3.3724
2	−0.0609	0.0531	−0.1375	3.3901	3.3412	3.5042
3	−0.0500	0.0320	−0.1320	3.4759	3.4264	3.3947
4	−0.0547	0.0930	−0.1859	3.3196	3.3575	3.5380
5	−0.0477	0.0672	−0.1656	3.3912	3.5331	3.3718

$I_{ab}(\mu)$, $I_{ca}(\mu)$, $I_{bc}(\mu)$ *are average values of the stator line currents healthy case.*
$I_{ab}(RMS)$, $I_{ca}(RMS)$, $I_{bc}(RMS)$ *are RMS values of the stator line currents healthy case.*

Table 3.
RMS and average values of the stator currents in healthy case.

fuzzy system is able to detect and classify the open-circuit condition by providing a I_{ca} value of 0.495, whereas the I_{ab} and I_{bc} values set at 0.165 for the healthy condition, as depicted in **Figure 20**.

Table 5 shows the RMS and average values of the line currents when 2 IGBT switches on a single branch, for the same line current, remains open because of a faulty condition. When this situation occurs, the corresponding line current, I_{ca} for this case, disappears, whereas the two remaining ones, I_{ab} and I_{bc}, locate at counterphase from each other with the same magnitude, as depicted in **Figure 8**. The obtained RMS values are utilized as input to the fuzzy-logic algorithm, producing

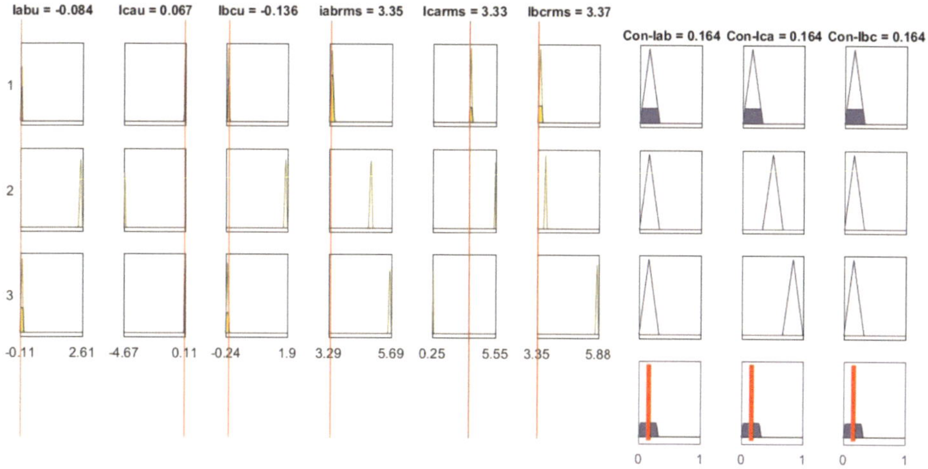

Figure 19.
Healthy case response throughout fuzzy logic.

Trial	$I_{ab}(\mu)$	$I_{ca}(\mu)$	$I_{bc}(\mu)$	$I_{ab}(RMS)$	$I_{ca}(RMS)$	$I_{bc}(RMS)$
1	2.4578	−4.5609	1.8867	4.7893	5.4575	3.7346
2	2.5922	−4.6555	1.8438	4.9517	5.5394	3.6856
3	2.5719	−4.5977	1.8117	4.8667	5.4884	3.6672
4	2.5937	−4.5930	1.7867	4.9066	5.5094	3.6758
5	2.4937	−4.5680	1.8672	4.8074	5.4729	3.6319

$I_{ab}(\mu)$, $I_{ca}(\mu)$, $I_{bc}(\mu)$ *are average values of the stator line currents positive semi-cycle family case.*
$I_{ab}(RMS)$, $I_{ca}(RMS)$, $I_{bc}(RMS)$ *are RMS values of the stator line currents healthy case positive semi-cycle family case.*

Table 4.
RMS and average values of the stator line currents with the positive semi-cycle switch in open-circuit fault condition in drive's phase A.

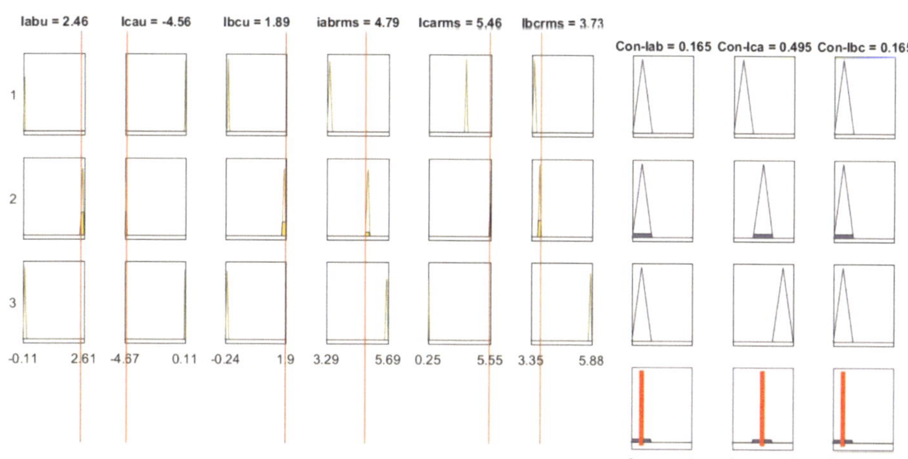

Figure 20.
Detection and isolation of one switch faulted corresponding to I_{ca} *current using fuzzy logic.*

Trial	$I_{ab}(\mu)$	$I_{ca}(\mu)$	$I_{bc}(\mu)$	$I_{ab}(RMS)$	$I_{ca}(RMS)$	$I_{bc}(RMS)$
1	0.0375	0.0430	−0.2211	5.6210	0.3061	5.8062
2	0.0078	0.0453	−0.1858	5.6220	0.2759	5.7950
3	0.0070	0.0680	−0.2195	5.6727	0.3073	5.8139
4	−0.0094	0.0563	−0.1734	5.5877	0.2910	5.7680
5	−0.0906	0.0969	−0.1461	5.6730	0.3257	5.8596

$I_{ab}(\mu)$, $I_{ca}(\mu)$, $I_{bc}(\mu)$ *are average values of the stator line currents both semi-cycles family case.*
$I_{ab}(RMS)$, $I_{ca}(RMS)$, $I_{bc}(RMS)$ *are RMS values of the stator line currents both semi-cycles family case.*

Table 5.
RMS and average values of the stator line currents with both semi-cycle switches in open-circuit fault condition in drive's phase A.

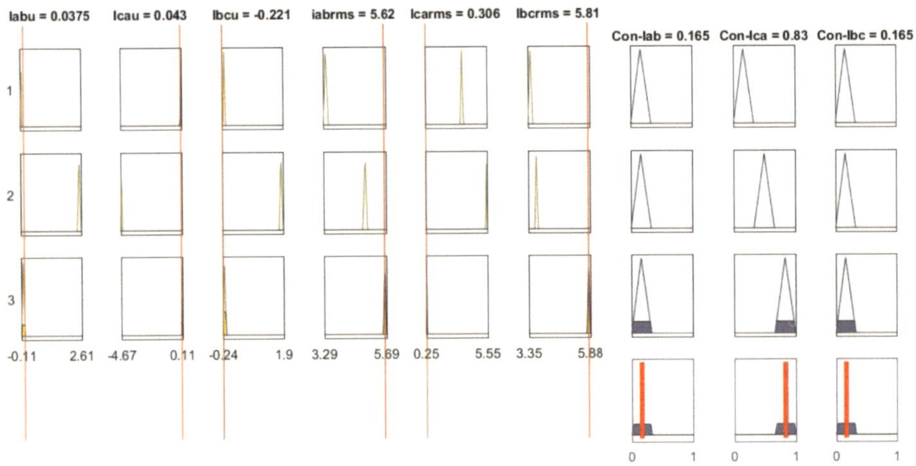

Figure 21.
Detection and isolation of two switches faulted corresponding to I_{ca} current using fuzzy logic.

an output value of 0.83 for I_{ca} in open-circuit condition and output values of 0.165 for I_{ab} and I_{bc} in healthy state, as shown in **Figure 21**.

7. Conclusions

This research presents some techniques for the detection and isolation of open-circuit faults in the IGBT technology switches of the power converter applied to a three-phase induction motor. One of the techniques used is the time-decimated base-2 discrete Fourier transform. This algorithm is fed with the mapping of the induction motor's stator currents to the alpha-beta reference frame, known as the Concordia transform, which uses the d-q transform or Park's instantaneous spatial vector (ISV). The main technique that allows fault detection and isolation is the phasor analysis of the line currents. A classifier based on fuzzy logic with the Mamdani algorithm is implemented to validate the fault isolation, which allows the identification of the faulty switch as well as the phase.

Acknowledgements

This work was partially supported by DEPI-ITA-TecNm (Departamento de Posgrado e Investigacion—Instituto Tecnologico de Aguascalientes/Tecnologico Nacional de Mexico), Departamento de Produccion y Seguridad Industrial, Universidad Tecnologica de Aguascalientes, Aguascalientes, Multidisciplinary Studies Department, Engineering Division, Campus Irapuato-Salamanca, University of Guanajuato.

Notes/thanks/other declarations

Thanks to the Consejo Nacional de Humanidades Ciencias y Tecnologias (CONAHCYT) for the Doctoral program scholarship.

Abbreviations

IM	induction motor
IMD	induction motor drive
MCSA	motor current signature analysis
MRA	multiresolution analysis
DSP	digital signal processor
PM	permanent magnet
IGBT	isolated gate bipolar transistor
ISV	instantaneous space vector
DC	direct current
PWM	pulse width modulation
FFT	Fast Fourier Transform
CPU	central processing unit

Nomenclature

$\mathbf{I}_{ab}$	Phase A to B line current
$\mathbf{I}_{bc}$	Phase B to C line current
$\mathbf{I}_{ca}$	Phase C to A line current
i_{ab}	Phase A to B instantaneous current
i_{bc}	Phase B to C instantaneous current
i_{ca}	Phase C to A instantaneous current
F_{stator}	stator fault frequency
F_{fault}	driver fault frequency
F_L	line frequency
I_α	α current in the α-β reference frame
I_β	β current in the α-β reference frame
I_0	0 current in the dq reference frame and α-β reference frame

Author details

Francisco Javier Villalobos-Pina[1*†], Josué A. Reyes-Malanche[2†], Eduardo Cabal-Yepez[3†] and Efrain Ramirez-Velasco[1†]

1 Departamento de Ingeniería Electrica Electronica, TecNm/Instituto Tecnologico de Aguascalientes, Aguascalientes, Mexico

2 Departamento de Produccion y Seguridad Industrial, Universidad Tecnologica de Aguascalientes, Aguascalientes, Mexico

3 Multidisciplinary Studies Department, Engineering Division, Campus Irapuato-Salamanca, University of Guanajuato, Guanajuato, Mexico

*Address all correspondence to: francisco.vp@aguascalientes.tecnm.mx; fvillalobospia@yahoo.com

† These authors contributed equally.

References

[1] Benslimane T. A new technique for simultaneous detection of one to two open-switch faults in three phase voltage-inverter-fed PM brushless DC motor drive. Journal of Electrical Engineering. 2008;**59**(2):97-100. DOI: 10.3390/en15217855

[2] Byoung-Gun P, Lee K-J, Kim R-Y, Hyun D-S. Low-cost fault diagnosis algorithm for switch open-damage in BLDC motor drives. Journal of Power Electronics. 2010;**10**(6):702-708. DOI: 10.6113/JPE.2010.10.6.702

[3] Wang Y, Li Z, Minghui X, Ma H. A comparative study of two diagnostic methods based on the switching voltage pattern for IGBT open-circuit faults in voltage-source inverters. Journal of Power Electronics. 2016;**16**(3):1087-1096. DOI: 10.6113/JPE.2016.16.3.1087

[4] Trabelsi M, Boussak M, Benbouzid M. Multiple criteria for high performance real-time diagnostic of singleand multiple open-switch faults in ac-motor drives: Application to IGBT-based voltage source inverter. Electric Power Systems Research. 2017;**2016**(144):136-149. DOI: 10.1016/j.epsr.2016.11.021

[5] Morel C, Rivero S, Le Gueux B, Portal J, Chahba S. Currents analysis of a brushless motor with inverter faults—Part I: Parameters of entropy functions and open-circuit faults detection. Actuators. 2023;**12**(6):228. DOI: 10.3390/act12060228

[6] Milanez DL, Emanuel AE. The instantaneous-space-phasor a powerful diagnosis tool. IEEE Transactions on Instrumentation and Measurement. 2003;**51**(1):143-148

[7] Cruz SMA, Cardoso AJM. Stator winding fault diagnosis in three-phase synchronous and asynchronous motors, by the extended Park's vector approach. IEEE Transactions on Industry Applications. 2001;**37**(5):1227-1233

[8] Reyes-Malanche JA, Villalobos-Pina FJ, Ramırez-Velasco E, Cabal-Yepez E, Hernandez-Gomez G, Lopez-RamirezCruz M. Short-circuit fault diagnosis on induction motors through electric current phasor analysis and fuzzy logic. MDPI Energies. 2023;**16**(516):1-15. DOI: 10.3390/en16010516 [Accessed: January 03, 2023]

[9] Walker JS. A Primer on Wavelets and their Scientific Applications. 2nd ed. USA: Chapman & Hall/CRC; 2008

Chapter 3

Implementing ADRC and PI Controls for a Sensorless Induction Motor Supplied by a Multi-Level Converter and Using PV Sources

Abdellah Oukassi

Abstract

The proposed book will focus on the controls of induction machines aimed at exploiting them in the field of speed variation (electric vehicles, aeronautics, etc.). In this work, the authors utilize cascaded multi-level converters to reduce the harmonic distortion rate of the voltages at the machine's input. The converter is controlled by pulse width modulation (PWM) technique. The use of active disturbance rejection control (ADRC) will contribute to eliminating internal disturbances (undesired variations in machine resistances) as well as minimizing external effects (inertia, resisting torque). State observers are employed to estimate mechanical speed, allowing the elimination of the speed sensor in the system. The mathematical models obtained have been simulated using MATLAB/Simulink software. Several scenarios have been tested through simulation, including comparing ADRC and PI control, and the effect of varying atmospheric conditions on PV power supply and consequently on the machine.

Keywords: induction motor, photovoltaic generator, multi-level converter, active disturbance rejection control, indirect field-oriented control, control PID, state observers

1. Introduction

In recent years, the integration of renewable energy sources with advanced control techniques has sparked growing interest in the field of electrical systems and industrial applications. Such an integrated system involves the combination of an induction machine, indirect vector control, power supply from photovoltaic (PV) panels, cascaded multilevel converters, and closed-loop control using either active disturbance rejection control (ADRC) or proportional-integral (PI) control. This integrated system represents a sophisticated solution aimed at effectively harnessing renewable energy while ensuring reliable and precise control of the induction machine's operation. By combining intermittent energy generation from PV panels with the dynamic performance of the induction machine, this system provides a sustainable energy solution suitable for various applications, ranging from water pumping to industrial processes.

Induction machines, also known as asynchronous motors, are widely used in variable-speed drives due to their excellent performance and reliability. Here are some key performances of the induction motors used in variable speed drives [1]. They operate stably over a wide load range and exhibit good tolerance to load variations, ensuring reliable operation in various conditions. They allow continuous speed adjustment over a wide range, making them suitable for a variety of applications, from fans to industrial conveyors. Compared to other types of motors, induction machines are often less expensive to manufacture and maintain, making them an attractive choice for many applications. Currently, vector control (direct or indirect) based on a mathematical model of the machine is widely used to achieve the previously mentioned performances [2, 3].

The use of renewable energies, especially solar energy, in electromechanical systems has grown considerably. Indeed, the performance of photovoltaic (PV) sources depends largely on the ability to maximize the capture and use of available solar energy. A crucial method to achieve this is the use of maximum power point tracking (MPPT) control [4]. MPPT systems optimize the performance of solar panels by continuously adjusting their operation to extract maximum energy possible, despite variations in environmental conditions such as sunlight and temperature. In the future, the focus will be on the use of photovoltaic sources to power static energy converters. These converters can be integrated into various real systems, such as electric cars, railways, and aerospace applications. Here, the focus is on the development of a variable-speed drive based on an induction machine and powered by photovoltaic sources.

Multilevel converters are increasingly used in electric vehicles due to their advantages in terms of energy efficiency and performance. They facilitate efficient conversion of DC power from PV panels into high-quality AC power, minimizing losses and harmonic distortions. Due to their design, multilevel converters can reduce electromagnetic interference [5], thereby improving the reliability and durability of electronic systems in electric vehicles. Their use in this context enables better management of electrical energy conversion, resulting in improved overall efficiency of the electric propulsion system in vehicles. Additionally, their use offers better control capability of electrical power, allowing for more precise regulation of the speed and torque of the electric motor.

The use of indirect vector control allows for precise control of the speed and torque of the induction machine, ensuring optimal performance in variable operating conditions. This control method is based on a cascade regulation loop that regulates both stator current and magnetic flux [6]. By adjusting these parameters, indirect vector control can accurately control the speed and torque of the induction machine, enabling fast dynamic response and excellent regulation accuracy. The use of indirect vector control offers numerous advantages, including improved energy efficiency, rapid response to load changes, and reduced mechanical stress on the machine. The parameters of the vector control will serve as reference signals, which will then drive the PWM control circuit of the multi-level converter.

Furthermore, the closed-loop control strategy, implemented either with ADRC or PI control, adds an extra layer of stability and robustness to the system. ADRC offers active disturbance rejection capabilities, allowing the system to adapt to changing conditions and effectively reject external disturbances [7]. This maintains high control performance even in the presence of external disturbances or load variations. Unlike some other advanced control methods, ADRC does not require a precise model of the system, simplifying its implementation in real-world applications where modeling may be difficult. Due to its ability to actively reject disturbances, ADRC can offer better reference tracking and disturbance rejection performance compared to PID control in some cases. On the other hand, PI control, a well-established method, offers simplicity and reliability, making it suitable for applications where precise modeling may be difficult [8, 9].

The use of state observers to estimate the mechanical speed of an induction machine is a crucial area in the control and monitoring of electromechanical systems. However, their mechanical speed, which is often a vital parameter for precise system control and monitoring, is not directly measurable. This is where state observers come into play. These sophisticated algorithms leverage available measurements, such as electrical currents and voltages, to estimate internal system variables like mechanical speed, which are not directly accessible. State observers are particularly useful in applications where dedicated sensors for certain variables are expensive or difficult to install. In this context, the Leunberger observer is an effective and robust method for estimating the mechanical speed of induction machines by utilizing available electrical measurements and combining them with a dynamic model of the system [10]. This approach plays a crucial role in the precise control and monitoring of these critical electromechanical systems in various industrial applications. The Leunberger observer offers several advantages, including its robustness against disturbances and model uncertainties, as well as its ability to operate effectively even under nonlinear operating conditions [11]. Additionally, it is relatively simple to implement and does not require detailed knowledge of the distribution of measurement errors, making it attractive for many industrial applications.

2. Modeling of the motor

Asynchronous machines, or induction machines, are among the most widespread and essential electromechanical devices in modern industry. Electrically, an asynchronous machine can be represented by a set of equations delineating the interactions among magnetic fields, currents, and voltages. And from an electromechanical perspective, the equations describe the relationship between electrical and mechanical quantities, such as electromagnetic torque and rotational speed. Here are the fundamental equations used in the mathematical model of an asynchronous machine in the form of a state-space model (Eq. (1)).

$$\begin{aligned} \frac{dx}{dt} &= [A][x] + [B][U] \\ [y] &= [C][x] \end{aligned} \tag{1}$$

With: $[x]^t = \begin{bmatrix} i_{sd} & i_{sq} & \phi_{rd} & \phi_{rq} \end{bmatrix}$, $[u]^t = \begin{bmatrix} v_{sd} & v_{sq} \end{bmatrix}$.

$$[A]^t = \begin{bmatrix} a_{11} & a_{12}\omega_s & a_{13} & a_{14}\omega_r \\ a_{21}\omega_s & a_{22} & a_{23}\omega_r & a_{24} \\ a_{31} & a_{32} & a_{33} & a_{34}\omega_g \\ a_{41} & a_{42} & a_{43}\omega_g & a_{44} \end{bmatrix} [B] = \begin{bmatrix} b_{11} & b_{12} \\ b_{21} & b_{22} \\ b_{31} & b_{32} \\ b_{41} & b_{42} \end{bmatrix}$$

$$a_{11} = a_{22} = -\left(\frac{1}{T_s\sigma} + \frac{1-\sigma}{T_r\sigma}\right), a_{12} = -a_{21} = 1, a_{13} = a_{24} = \frac{1-\sigma}{T_r M\sigma},$$
$$a_{14} = -a_{23} = \frac{1-\sigma}{\sigma M} a_{32} = a_{41} = 0, a_{33} = a_{44} = \frac{-1}{T_r}, a_{34} = -a_{43} = 1,$$
$$b_{11} = b_{22} = \frac{1}{\sigma \cdot L_s}, b_{31} = b_{32} = b_{41} = b_{42} = b_{12} = b_{21} = 0$$

The coefficients that may be affected by the variation of the machine parameters (stator resistance and rotor resistance) are: $a_{11}, a_{13}, a_{42}, a_{44}$. The matrix$[A']$can be written as (Eq. (2)).

$$[A'] = [A] + [\Delta A] \tag{2}$$

With: $[\Delta A] = \begin{bmatrix} \Delta a_{11} & 0 & \Delta a_{13} & 0 \\ 0 & \Delta a_{22} & 0 & \Delta a_{24} \\ \Delta a_{31} & 0 & \Delta a_{33} & 0 \\ 0 & \Delta a_{42} & 0 & \Delta a_{44} \end{bmatrix}$ $\Delta a_{11} = \Delta a_{22} = -\left(\frac{K_{rs}}{T_s\sigma} + K_{rr}\frac{1-\sigma}{T_r\sigma}\right)$,
$\Delta a_{13} = \Delta a_{24} = K_{rr}\frac{1-\sigma}{T_r M\sigma}$, $\Delta a_{31} = \Delta a_{42} = K_{rr}\frac{M}{T_r}\Delta a_{31} = \Delta a_{42} = K_{rr}\frac{M}{T_r}$,
$\Delta a_{33} = \Delta a_{44} = K_{rr}\frac{-1}{T_r}$, $0 \le K_{rs}\% \le 100$, $0 \le K_{rr}\% \le 100$. K_{rr} and K_{rs} are the coefficients of resistance variation.

3. Pv source

Photovoltaic sources are widely used in various applications, including residential, commercial, and industrial electricity production, standalone power generation systems, large-scale solar power plants, etc. In terms of equations, mathematical models of photovoltaic sources can be based on physical laws such as the current-voltage characteristic equation, equations describing temperature-dependent behavior, and equations describing efficiency variation with solar irradiance [12]. These equations allow predicting the electricity production of photovoltaic sources under different environmental and operational conditions. In our case, these sources are used to power the multi-level converter. The fundamental equations used to model a photovoltaic cell include:

The current generated by the panel can be given as (Eq. (3)):

$$I = I_{ph} - I_D - I_P \tag{3}$$

The relationship between I and V is given by Eq. (4):

$$I = I_{ph} - I_0\left(\exp\left(\frac{V + R_s I}{a.Vt}\right) - 1\right) - \frac{V + R_s.I}{R_p} \tag{4}$$

The Iph is a grandeur that gives information on the spectrum of the photovoltaic cell. Practically, its evolution is related to ambient temperature and irradiation G. Its formula is Eq. (5):

$$I_{ph} = \frac{G}{Gr}(Iscr - Ki(T - Tr)) \tag{5}$$

Symbol	Parameter	Value
K	Boltzmann Constant	1.38065e-23 j/K
I_{scn}	Nominal SC Current	8.21 A
V_{ocn}	Nominal OC voltage	32.9 V
K_v	Temperature Voltage constant	0.123 K
K_i	Temperature current constant	0.003 K
N_s	Number of series connected cells	48
R_s	Series Resistor	0.221 Ω
R_{sh}	Shunt Resistor	415.405 Ω
T	Nominal temperature	298 K

Table 1.
PV device specifications (1 kW/m², 25°C).

T_r is the reference temperature, I_{scr} is the cell's short circuit current at T_r, k_i is the temperature coefficient of the short circuit, and G_r is the nominal irradiation. The diode saturation current I_0 depending on temperature may be expressed by Eq. (6):

$$I_0 = I_{or}\left(\frac{T}{Tr}\right)^3 . \exp\left(\frac{qEg}{Ak}\left(\frac{1}{Tr} - \frac{1}{T}\right)\right) \tag{6}$$

Where I_{or} is the nominal saturation current, and E_g is the band gap energy. Usually, a PV installation is broken down into several cells (series or parallel). Before operating a cell, it is necessary to know its physical parameters (e.g., **Table 1** illustrates the characteristics of a panel).

4. Multi-level inverter modeling with PWM control

4.1 Topology

Multilevel converters have emerged as a crucial technology in modern power electronics, offering advantages such as higher voltage levels, reduced harmonic distortion, and improved efficiency compared to traditional two-level converters. Several types of multilevel converters exist, each with its own unique configuration and characteristics. One common type is the neutral-point clamped (NPC) converter [13], which features capacitors connected between the neutral point and each phase to achieve multiple voltage levels. Another type is the flying capacitor (FC) converter [14], which utilizes a set of capacitors connected across the semiconductor switches to generate the desired voltage levels. Additionally, the cascaded H-bridge (CHB) converter is widely used in high-power applications, employing multiple H-bridge cells stacked in series to achieve multiple voltage levels. Other variants include the modular multilevel converter (MMC), which utilizes submodules interconnected in a series-parallel configuration, offering scalability and fault tolerance. The cascaded H-bridge (CHB) inverter topology consists of multiple H-bridge inverter modules connected in series [15]. This configuration enables the inverter to generate a higher number of

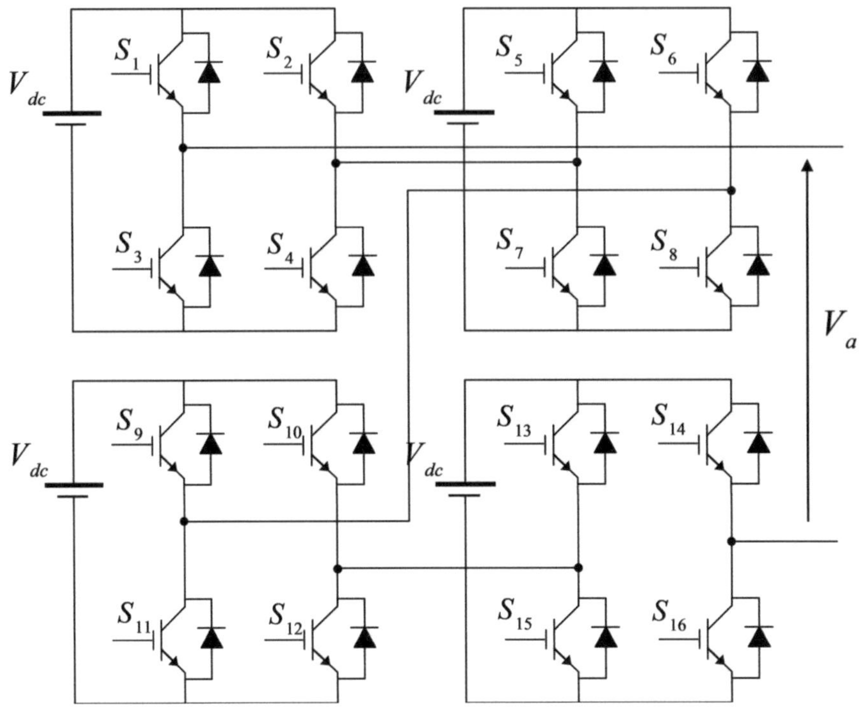

Figure 1.
Converter structure (nine CHBs) for a single phase.

voltage levels, thus improving the quality of the output waveform. In this configuration (**Figure 1**), the converter consists of 12 single-phase H-bridge modules arranged to power a three-phase induction machine. Each phase of the induction machine is powered by 4 single-phase H-bridge modules. Each H-bridge module can switch between three voltage states: V_{dc}, 0, and $-V_{dc}$.

4.2 PWM control

The basic principle of the multi-carrier MCPWM is the use of multiple triangular carrier signals of different amplitudes or phases, which are then compared to a single sine waveform to generate the control pulses. The best known MCPWM classes are: phase disposition (PD), phase opposed disposition (POD), and alternative phase opposed disposition (APOD). In the technique (PD), all carriers are in phases. In the POD technique, carriers are all in phase both above and below the zero-reference value. Nonetheless, there exists a 180-degree phase shift between those above and below zero. Conversely, in the technique (APOD), carriers are alternately two-phase by 180 degrees in relation to each other [16]. A modified PWM technique to improve the total harmonic distortion of the multilevel inverter is used [17]. **Table 1** shows the algorithmic organization of the proposed PWM control to drive the switches of each bridge. The advantage of this structure is to eliminate the logic gates found in other structures. **Table 1** clearly indicates each phase: the parameters of the sinusoidal voltage source (peak and frequency), the parameters of each triangular wave (frequency of each triangular wave and amplitude variation interval), and the four bridges (switches used in each bridge). The proposed algorithm allows that for each bridge:

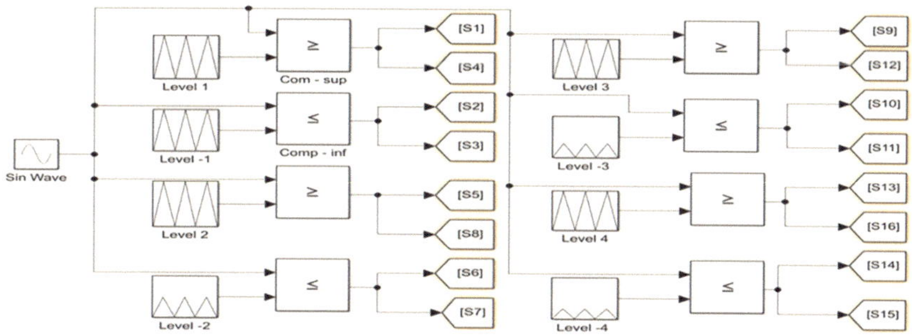

Figure 2.
PWM control electronic circuit to generate pulses for 16 IGBT.

Sine wave	Triangular wavef = 2 KHz		Comparison between sine wave and triangular waves		H-Bridge	H-Bridge switches
	Level	Volts	Comp-sup	Comp-inf		
Peak 1 Volts f = 50 Hz	+1	[0 0.25]	Sin wave > level + 1		1	(S1-S4)
	−1	[-0.25 0]		Sin wave < level-1	1	(S2-S3)
	+2	[0.25 0.5]	Sin wave > level + 2		2	(S5-S8)
	−2	[−0.5–.25]		Sin wave < level-2	2	(S6-S7)
	+3	[0.5 0.75]	Sin wave > level + 3		3	(S9-S12)
	−3	[−0.75–.5]		Sin wave < level-3	3	(S10-S11)
	+4	[0.75 1]	Sin wave > level + 4		4	(S13-S16)
	−4	[−1–0.75]		Sin wave < level-4	4	(S14-S15)

Table 2.
Organization of the PWM control logic for a 9-level converter.

A sine wave is compared (Comp sup) to a triangular wave having a positive level. If the sine wave is greater than the positive triangular wave, then pulses are sent to two opposite switches (e.g., S1 and S4).

The same sine wave is compared (Comp inf) to a negative triangular wave having the same level as the positive triangular wave previously used. If the sine wave is less than the negative triangular wave, pulses are sent to the remaining two switches (e.g., S2 and S3). **Figure 2** shows the logic control circuit proposed in **Table 2**.

5. Indirect field-oriented control approach

Indirect field-oriented control (IFOC) is a sophisticated control approach widely employed in the field of electric drives, particularly for AC induction motors and permanent magnet synchronous motors (PMSMs). This method is designed to achieve precise control of motor speed and torque while minimizing losses and improving efficiency. At its core, IFOC aims to decouple the motor's electromagnetic dynamics, separating the torque and flux components [1]. By utilizing the relationships between magnetic fluxes and currents in the rotor windings (Eqs. (7) and (8)):

$$0 = R_r i_{rd} + \frac{d\varphi_{rd}}{dt} - \omega_g \varphi_{rq} \tag{7}$$

$$0 = R_r i_{rd} + \frac{d\varphi_{rq}}{dt} + \omega_g \varphi_{rd} \tag{8}$$

For decoupling control, we have $\varphi_{rq} = 0$, $\varphi_{rd} = \varphi_r$. The expressions of the sliding pulsation (Eq. (9)) and the torque (Eq. (10)) are:

$$\hat{\omega}_g = \frac{i_{sq}}{i_\varphi L_r} \hat{R}_r \tag{9}$$

$$C_e = p \frac{M^2}{L_r} i_\varphi i_{sq} \tag{10}$$

The mechanical speed estimated as a function of electromagnetic torque and load torque is given by the following expression (Eq. (11)).

$$J \frac{d\hat{\Omega}_r}{dt} = -K_f \hat{\Omega}_r + (C_e - C_r) \tag{11}$$

An additional set of expressions, utilizing the concept of decoupling within the framework of IFOC, are presented by Eq. (12). These equations describe the relationships between voltages and currents in the stator windings of the machine.

$$\begin{aligned} v_{sd}^* &= \sigma Ls \frac{di_{sd}}{dt} + \sigma Ls \left(\frac{1}{T_s \cdot \sigma} + \frac{1-\sigma}{T_r \cdot \sigma} \right) i^*{}_{sd} \\ v_{sq}^* &= \sigma Ls \frac{di_{sq}}{dt} + \sigma Ls \left(\frac{1}{T_s \cdot \sigma} + \frac{1-\sigma}{T_r \cdot \sigma} \right) i^*{}_{sq} \end{aligned} \tag{12}$$

6. ADRC control

ADRC stands for active disturbance rejection control, which is an advanced control strategy used in various engineering applications, particularly in the field of control systems. ADRC is designed to actively reject disturbances and uncertainties in the system by employing an extended state observer (ESO) to estimate and compensate for these disturbances in real time. ADRC is the result of proportional-integral derivation (PID), the key to its success, and this is a control based on errors rather than on a model [18]. **Figure 3** provides a visual representation of how these components are interconnected and how information flows within the ADRC system, illustrating the overall architecture and functionality of the control system.

This Equation (Eq. (13)) describes how the output y of the system evolves over time based on its internal dynamics (y, ε, t) and the input signal u, with b_0 representing the scaling factor or sensitivity of the system to the input [19].

$$y = f(y, \varepsilon, t) + b_0 u \tag{13}$$

Using the state-space form, equation (Eq. (13)) can be represented as follows (Eq. (14)):

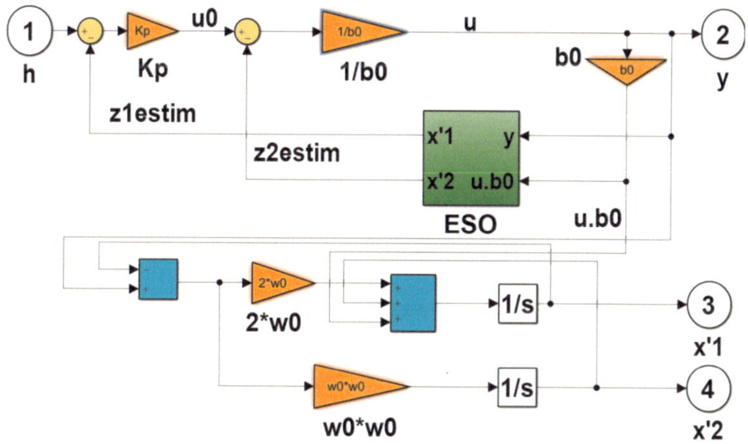

Figure 3.
Interconnection between components of the ADRC system.

$$\begin{cases} y = \begin{pmatrix} 1 \\ 0 \end{pmatrix}^T .x \\ \dot{x} = \begin{pmatrix} 0 & 1 \\ 0 & 0 \end{pmatrix}.x + \begin{pmatrix} 1 \\ 0 \end{pmatrix}.b_0 u + \begin{pmatrix} 0 \\ 1 \end{pmatrix}.\dot{f} \end{cases} \tag{14}$$

The extended state observer (ESO) is commonly represented by an equation that estimates the internal state of the system as well as the disturbances affecting the system dynamics. A generic form of the ESO equation could be (Eq. (15)).

$$\begin{cases} \hat{y} = \begin{pmatrix} 1 \\ 0 \end{pmatrix}^T .z \\ \dot{z} = \begin{pmatrix} 0 & 1 \\ 0 & 0 \end{pmatrix}.z + \begin{pmatrix} 1 \\ 0 \end{pmatrix}.b_0 u + \begin{pmatrix} 2\omega_0 \\ \omega_0^2 \end{pmatrix}.(y - \hat{y}) \end{cases} \tag{15}$$

with ω_0 the observer bandwidth, determined by pole placement, in order to provide a fast, high-quality output [20]. According to ADRC, a law of control is then determined by (Eq. (16)):

$$u = \frac{u_0 - z_2}{b_0} \tag{16}$$

With: z_2 being a correct estimation of "f" and z_1 a correct estimation of "y".
The resulting equation (Eq. (13)) becomes (Eq. (17)):

$$\begin{cases} u_0 = K_p(h - z_1) \\ \dot{y} = u_0 + (f - z_2) \approx u_0 \end{cases} \tag{17}$$

Where: h is the reference input signal.
The equations (Eqs. (11) and (12)) above can be written in ADRC canonical form as the corresponding equations below (Eqs. (18)–(20)).

$$\frac{di_{ds}}{dt} = f_d(i_{ds}, \varepsilon, t) + \frac{V_{ds}(t)}{L_s\sigma} \tag{18}$$

$$\frac{di_{qs}}{dt} = f_q(i_{qs}, \varepsilon, t) + \frac{V_{qs}(t)}{L_s\sigma} \tag{19}$$

$$\frac{d\Omega}{dt} = f_\Omega(\Omega, \varepsilon, t) + \frac{C_e(t)}{J} \tag{20}$$

7. Luenberger state observer

The Luenberger observer is a crucial tool in control theory used to estimate the state of a dynamic system based on its measured outputs. The classical formulation of the Luenberger observer is given by the following equation (Eq. (21)). In your context, it is indicated that the coefficients of the matrix D are chosen so that the observed model is faster than the real model. This means that the eigenvalues of the matrix $A-DC$ should be chosen such that they are larger in absolute value than those of the matrix A [21, 22].

$$\begin{cases} \hat{y} = A.\hat{x} \\ \dot{\hat{x}} = B.\hat{x} + Cu + D.(y - \hat{y}) \end{cases} \tag{21}$$

The real physical model of the machine is identical to the one found in Section 2. It simply requires linking the stator reference frame to the (α, β) reference frame. The state vectors, the different matrices of the machine, and the coefficients of the observation matrix are as follows:

$$A = \begin{bmatrix} -\frac{1}{\sigma}\left(\frac{1}{T_s} + \frac{(1-\sigma)R_r}{L_r}\right) & 0 & \frac{MR_r}{\sigma L_s L_r^2} & \frac{M\omega_r}{\sigma L_s L_r} \\ 0 & -\frac{1}{\sigma}\left(\frac{1}{T_s} + \frac{(1-\sigma)R_r}{L_r}\right) & \frac{M\omega_r}{\sigma L_s L_r} & \frac{MR_r}{\sigma L_s L_r^2} \\ \frac{MR_r}{L_r} & 0 & -\frac{R_r}{L_r} & -\omega_r \\ 0 & \frac{MR_r}{L_r} & \omega_r & -\frac{R_r}{L_r} \end{bmatrix}$$

$$B = \begin{bmatrix} \frac{1}{\sigma L_s} & 0 & 0 & 0 \\ 0 & \frac{1}{\sigma L_s} & 0 & 0 \end{bmatrix}, C = \begin{bmatrix} 1 & 0 & 0 & 0 \\ 0 & 1 & 0 & 0 \end{bmatrix}, D = \begin{bmatrix} d_{11} & d_{12} & d_{13} & d_{14} \\ -d_{12} & d_{11} & -d_{14} & d \end{bmatrix}$$

$$d_{11} = (k-1)\left(\frac{1}{\sigma}\left(\frac{1}{T_s} + \frac{(1-\sigma)R_r}{L_r}\right) + \frac{R_r}{L_r}\right)$$

$$d_{12} = (1-k)\omega_r$$

$$d_{13} = (1-k^2)\left(\frac{MR_r}{L_r} - \frac{L_sL_r}{M}\left(\frac{1}{T_s} + \frac{(1-\sigma)R_r}{L_r}\right)\right) + \frac{L_sL_r(1-k)}{M}\left(\frac{1}{\sigma}\left(\frac{1}{T_s} + \frac{(1-\sigma)R_r}{L_r}\right) + \frac{R_r}{L_r}\right)$$

$$d_{14} = \frac{\sigma L_s L_r \omega_r (1-k)}{M}$$

Lyapunov function for mechanical speed systems can be conceptualized to analyze the stability of the speed dynamics of the system. To estimate a mechanical speed (Eq. (22)), the Lyapunov function is utilized [23].

$$V = (x - \hat{x})(x - \hat{x})^T + \frac{(\omega_r - \widehat{\omega_r})^2}{\rho} \tag{22}$$

With: ρ being a positive constant coefficient. $x - \hat{x}$ is the estimation error vector of the induction motor and LO. The derivative of V in time, which should be negative, is defined by the following equation (Eq. (23)):

$$\frac{dV}{dt} = \frac{d(x - \hat{x})}{dt}(x - \hat{x})^T + \frac{(x - \hat{x})^T}{dt}(x - \hat{x}) + \frac{1}{\rho}\frac{d(\omega_r - \widehat{\omega_r})^2}{dt} \tag{23}$$

The equation (Eq. (23)) can be developed by (Eq. (24)):

$$\begin{aligned}\frac{dV}{dt} = &\left[(B - AD)^T + (B - AD)\right](x - \hat{x})^T(x - \hat{x}) + 2(\omega_r - \widehat{\omega_r}) \\ &\frac{\sigma L_s L_r \widehat{\omega_r}(1 - k)}{M}\left(\widehat{\varphi_{\beta r}}\left(i_{\alpha s} - \widehat{i_{\alpha s}}\right) - \widehat{\varphi_{\alpha r}}\left(i_{\beta s} - \widehat{i_{\beta s}}\right)\right) \\ &- \frac{2(\omega_r - \widehat{\omega_r})}{\rho}\frac{d\widehat{\omega_r}}{dt}\end{aligned} \tag{24}$$

The expression $\left[(B - AD)^T + (B - AD)\right](x - \hat{x})^T(x - \hat{x})$ is always negative. Both last terms may be set to zero since they are insignificant in comparison with the first (Eq. (25)).

$$2(\omega_r - \omega_r)\frac{\sigma L_s L_r \widehat{\omega_r}(1 - k)}{M}\left(\widehat{\varphi_{\beta r}}\left(i_{\alpha s} - \widehat{i_{\alpha s}}\right) - \widehat{\varphi_{\alpha r}}\left(i_{\beta s} - \widehat{i_{\beta s}}\right)\right)\left(-\frac{2(\omega_r - \widehat{\omega_r})}{\rho}\frac{d\widehat{\omega_r}}{dt} = 0 \tag{25}$$

Thus,

$$\widehat{\omega_r} = \rho\frac{\sigma L_s L_r \omega_r(1 - k)}{M}\int\left(\widehat{\varphi_{\beta r}}\left(i_{\alpha s} - \widehat{i_{\alpha s}}\right) - \widehat{\varphi_{\alpha r}}\left(i_{\beta s} - \widehat{i_{\beta s}}\right)\right)dt \tag{26}$$

Many scientists have suggested a PI adaptation mechanism for improving estimation accuracy. It gives the following expression (Eq. (27)) for the estimated speed:

$$\widehat{\omega_r} = \left(K_p + \frac{K_i}{s}\right)\left(\widehat{\varphi_{\beta r}}\left(i_{\alpha s} - \widehat{i_{\alpha s}}\right) - \widehat{\varphi_{\alpha r}}\left(i_{\beta s} - \widehat{i_{\beta s}}\right)\right) \tag{27}$$

Where K_p and K_i are positive constants.

8. System proposed

The proposed (ADRC-IFOC- PI) diagram for the sensorless induction motor fed by HCBI is depicted in **Figure 4**. This system comprises interconnected subsystems aimed at maximizing performance. In the power supply segment, photovoltaic generators supply energy to a 9-level inverter *via* a boost converter. The multilevel

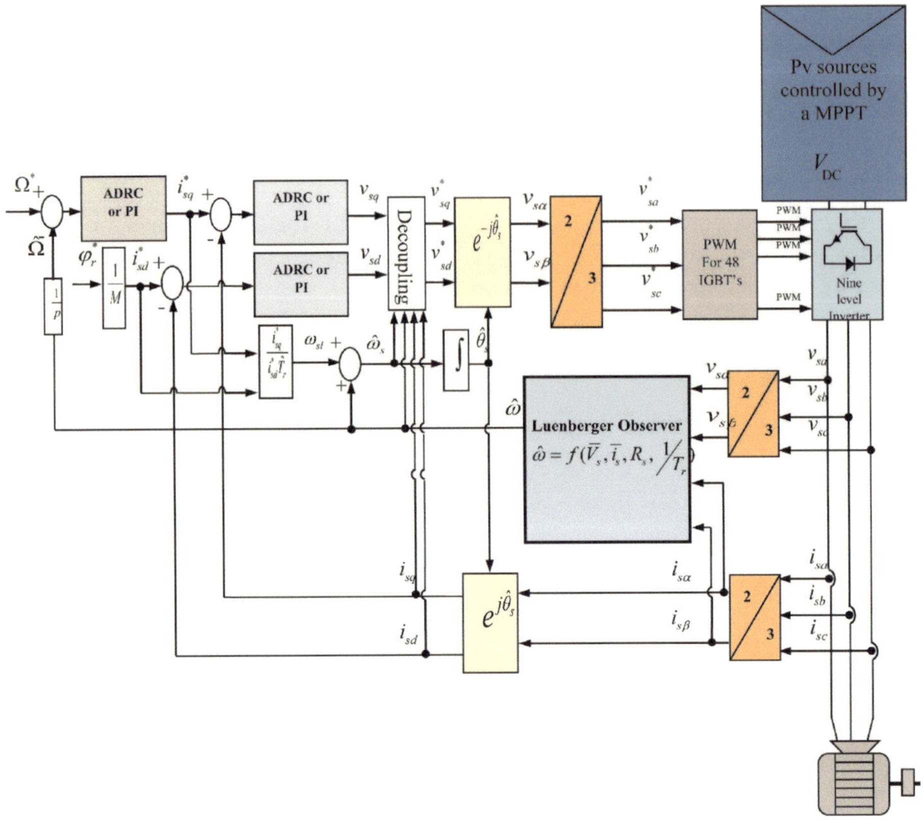

Figure 4.
Optimized operation of sensorless induction machine via *nine-level converter with ADRC and PI controls.*

inverter's output voltages are directly applied to the induction machine. A Luenberger observer is employed to estimate mechanical speed in the absence of sensors. Leveraging this observer, the estimated speed can be compared to a reference using a control law based on either the ADRC or PI approach. By integrating the indirect field-oriented control (IFOC) method with these two approaches, the system generates requisite signals for driving PWM control vectors, thus enhancing overall system efficiency and performance.

9. Results and discussions

The characteristics of the induction machine are illustrated in **Table 3**. Two scenarios were considered. The simulation tests conducted in the first scenario consist of not varying the parameters of the machine (stator and rotor resistances, moment of inertia). The second scenario involves considering these variations at different times.

9.1 Scenario 1: variation of the speed reference and load torque

The PV module is tested taking into account the values of the different given parameters and according to the conditions specified in **Table 1**(T = 25 degrees,

Parameters	Value
Power	1kw
Voltage	220/380 V
Current	4.58A/2.56A
Stator resistance	6,8 Ω
Rotor resistance Rr	5,43 Ω
Stator inductance Ls	0,3973 H
Rotor inductance Lr	0,3558 H
Mutual inductance M	0,39 H
Number of pole pairs	2
Viscous friction coefficient K_f	0,0025 Nm.s/rad
Rated speed n	1440Tr/min
Rotor time constant Tr	0.0655 s
Stator time constant Ts	0.0584 s

Table 3.
Parameters of induction machine.

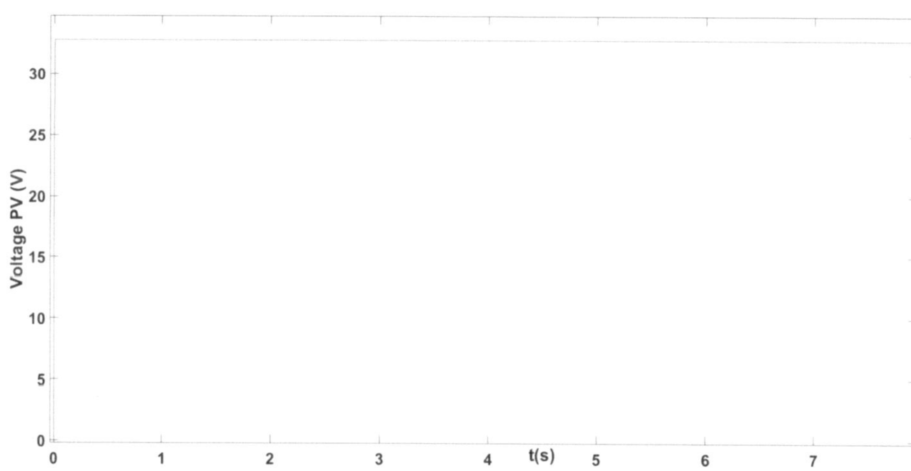

Figure 5.
Output voltage (Pv source).

I = 1000 W/m^2). The output voltage is illustrated in **Figure 5** (between 0 and 8 s). The PV module is tested taking into account the values of the different given parameters and according to the conditions specified in **Table 1**(T = 25, degrees I = 1000 W/m2). The output voltage is illustrated in **Figure 5** (between 0 and 8 s).

In this part, the simulation is performed by varying only the speed and load torque. In the interval [0, 5 s], the speed is equal to 90 rad/s, and in the interval [5 s, 8 s], the speed is equal to 140 rad/s. At t = 2 s, the torque goes from 0 to 5 Nm, and at t = 6 s, the torque goes from 5 Nm to 0 Nm. **Figure 6** shows the evolution of the three voltages obtained at the output of the nine-level converter, with a zoom on the

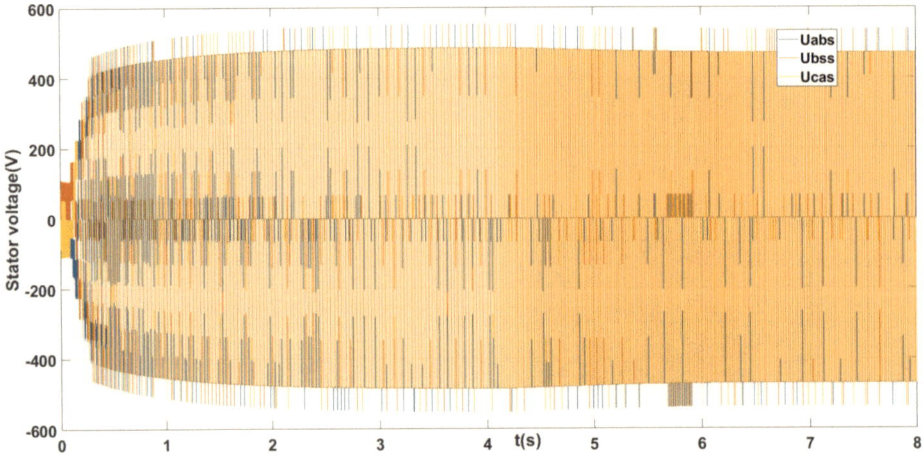

Figure 6.
Waveforms of line-to-line voltages of three phases (a,b,c).

interval [0.4 s, 0.5 s] illustrated in **Figure 7**. The stator currents are shown in **Figure 8**, and a zoom on the interval [0.4 s, 0.5 s] is provided in **Figure 9**. **Figure 10** presents the evolution of the mechanical speed in both cases (ADRC, PI), clearly demonstrating the faster response time and better performance of the ADRC control compared to the PI control.

Figure 11 shows the evolution of the electromagnetic torque in both scenarios (ADRC, PI). Additionally, the torque obtained with ADRC control exhibits sudden variations after a change in load torque, unlike the torque obtained with PI control.

To verify the validity of the state observer, rotor fluxes were also simulated. **Figure 12** illustrates the evolution of the two components of the flux along the α and β axes, and a zoom on the interval [0.4 s, 0.5 s] is provided in **Figure 13**.

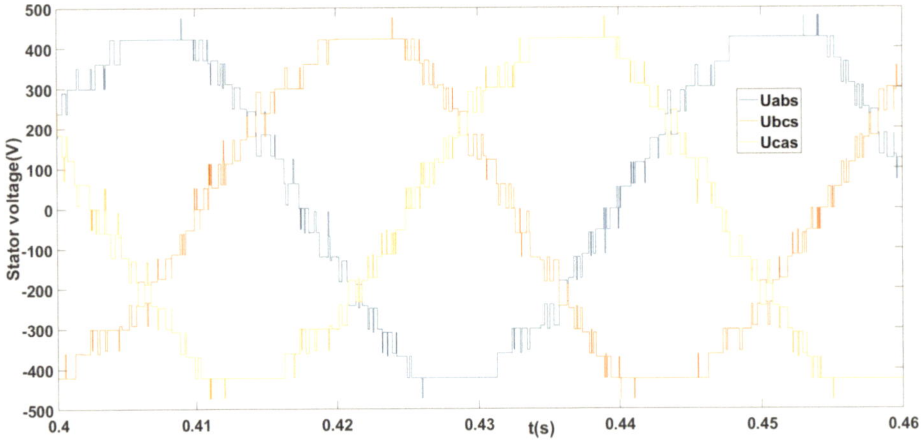

Figure 7.
Waveforms of line-to-line voltages loop (0.4 sec–0.46 sec).

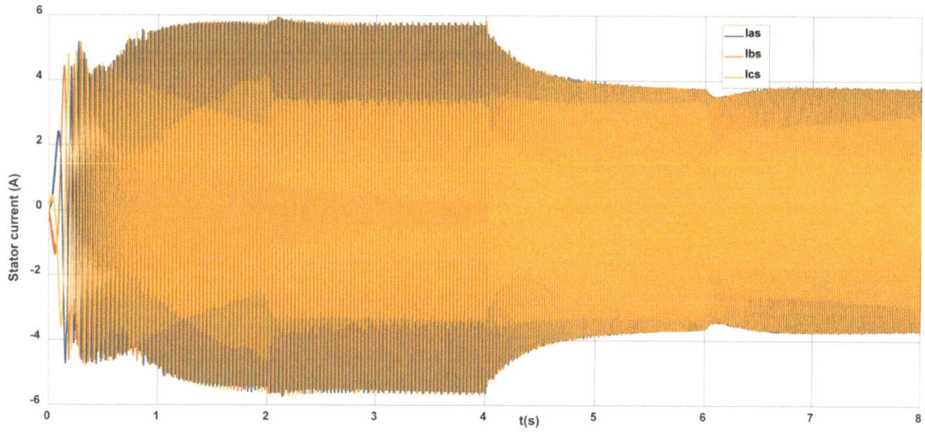

Figure 8.
Waveforms of stator currents.

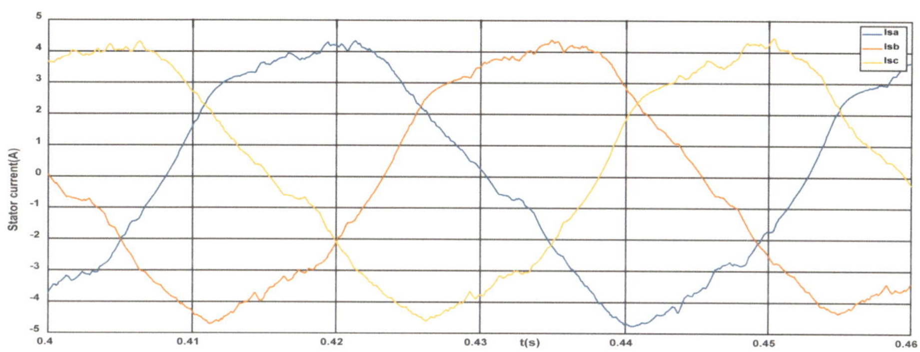

Figure 9.
Waveforms of stator currents loop (0.4 sec–0.46 sec).

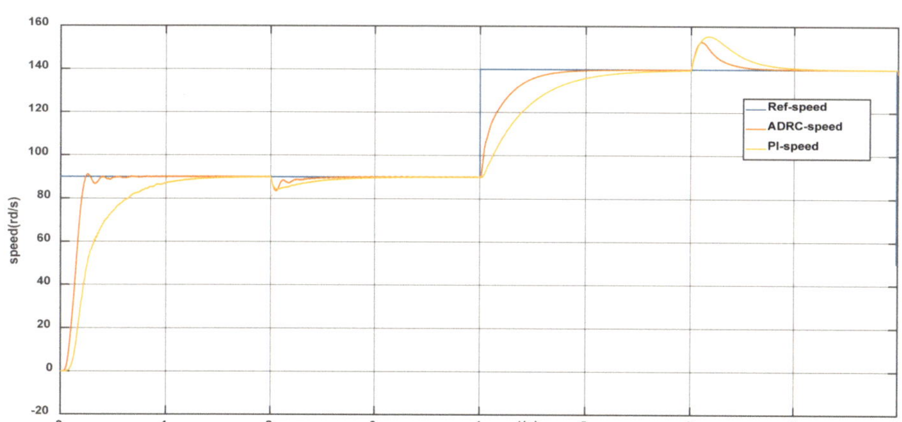

Figure 10.
Speed and reference speed.

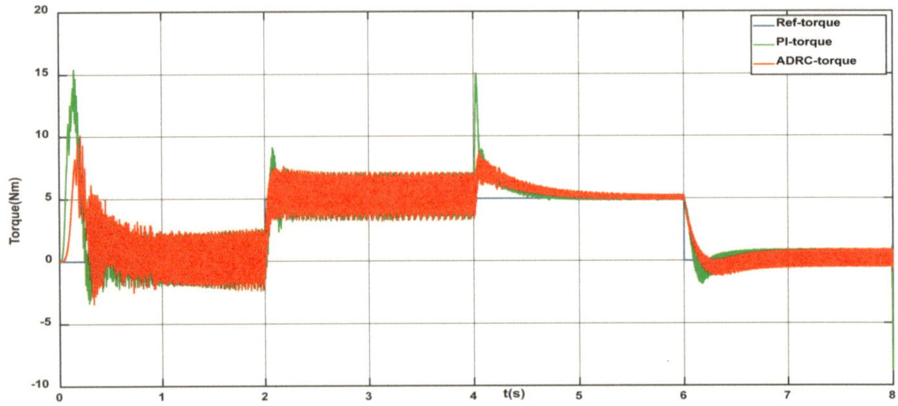

Figure 11.
Torque, load torque, and reference torque.

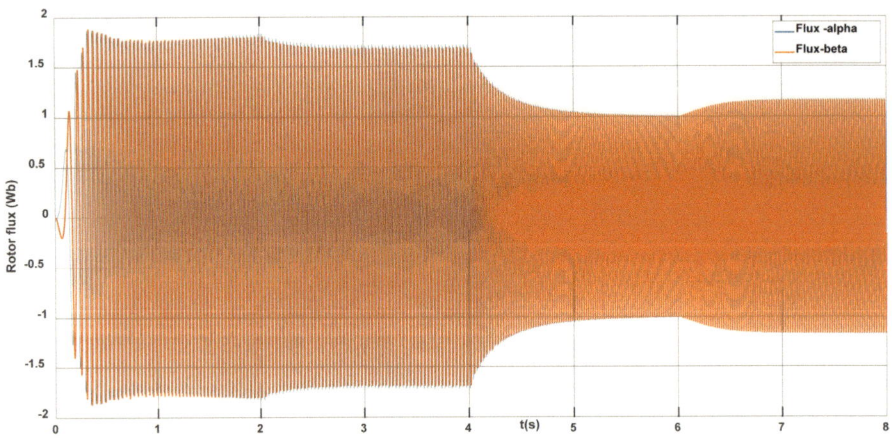

Figure 12.
Components of the flux along the α and β axes.

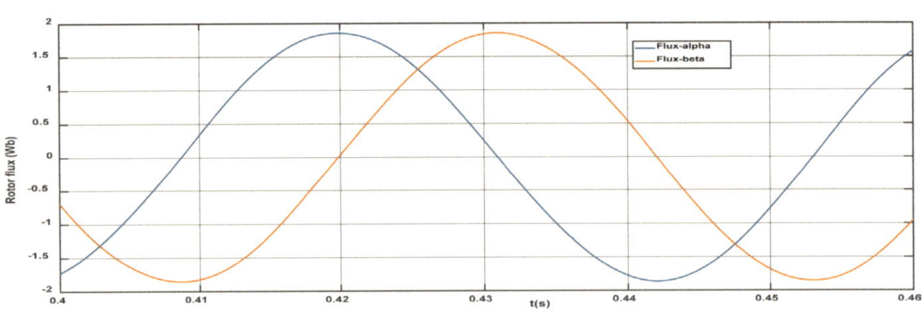

Figure 13.
Components of the flux along the α and β axes loop (0.4 sec – 0.46 sec).

9.2 Scenario 2: consideration of parameter variations (resistances and moment of inertia)

In the second part, the robustness of the proposed system is examined following variations in internal parameters (machine resistances) and external parameters (moment of inertia). The simulation tests were conducted with the following variations: For the machine, the rotor resistance was increased by 50% at t = 3.5 s, and the moment of inertia was also increased by 50% at t = 5 s. **Figure 14** illustrates the evolution of the mechanical speed (actual and estimated) in both cases (ADRC, PI), clearly indicating the faster response time and superior performance of ADRC control compared to PI control. Similarly, the error between the actual speed and the estimated speed in the ADRC system is lower than in the PI system. **Figure 15** shows the evolution of the

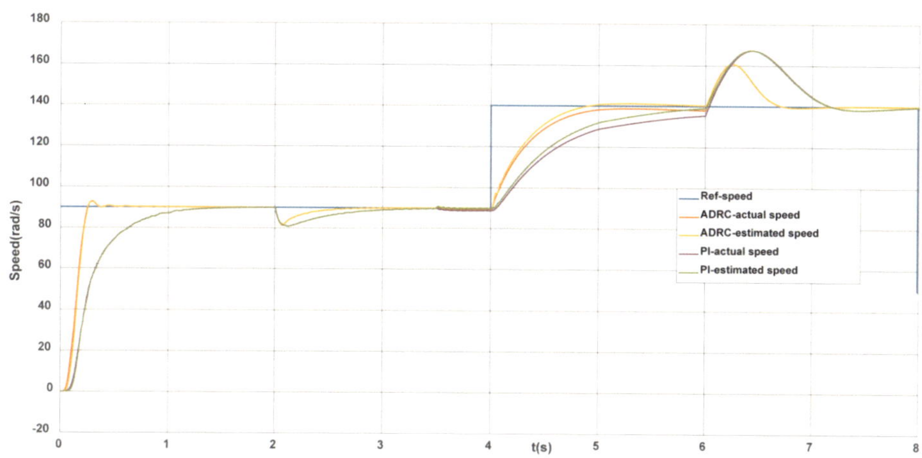

Figure 14.
Speed and reference speed.

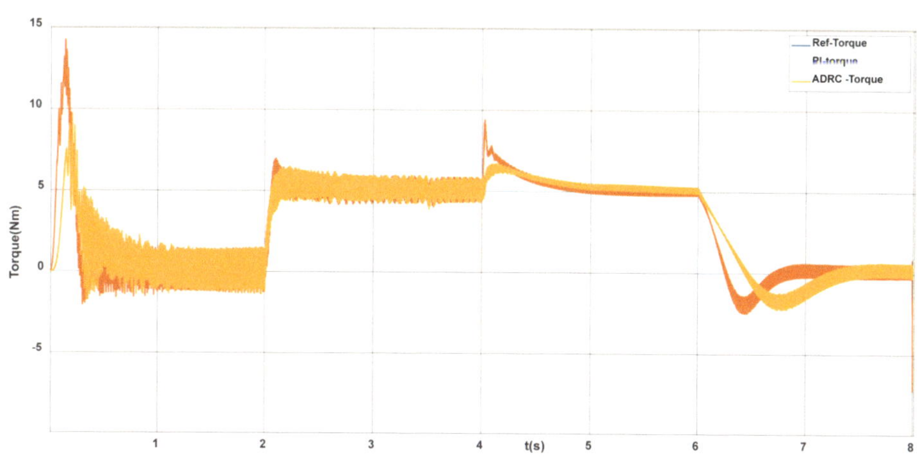

Figure 15.
Torque, load torque, and reference torque.

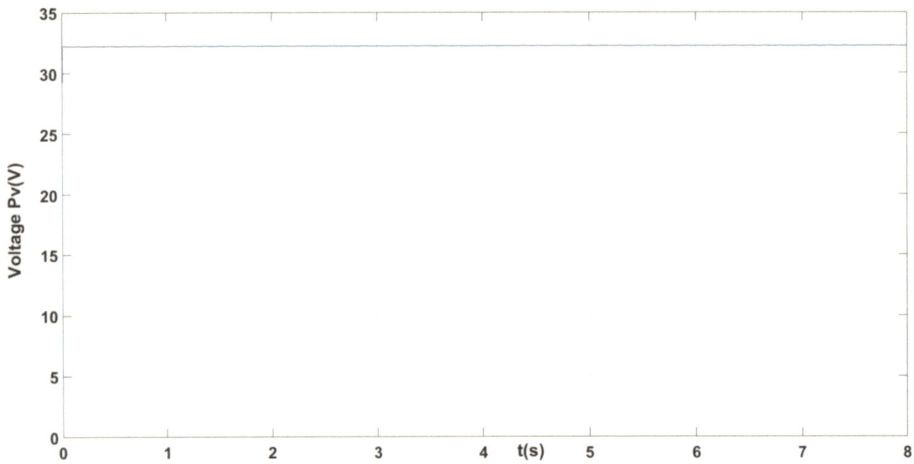

Figure 16.
Output voltage (PV source).

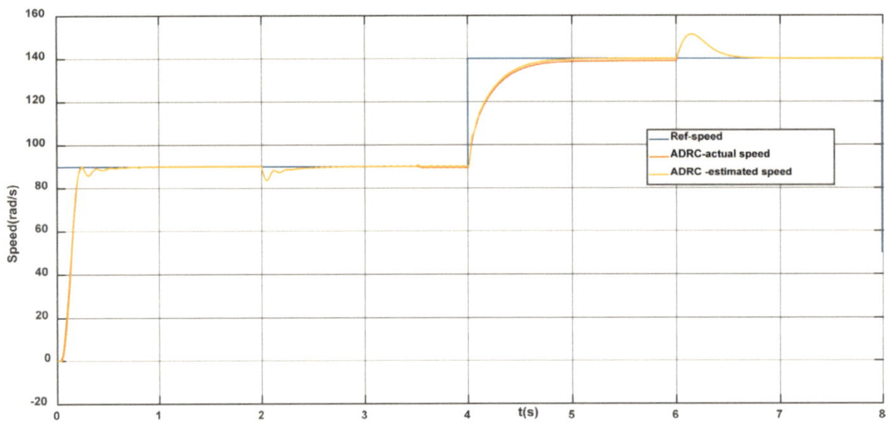

Figure 17.
Speed and reference speed.

electromagnetic torque in both cases (ADRC, PI). Additionally, the torque obtained with ADRC control exhibits lower interference than that obtained with PI control.

The simulated results at the end of this work involve varying the parameters of the PV sources (at t = 0, we take T = 35 degrees and G = 1100 W/m2). We still retain the variations of the machine parameters mentioned above. **Figure 16** depicts the profile of the voltage at the output of the PV module. A slight difference can be noticed compared to the one obtained in **Figure 5**. **Figure 17** shows the evolution of the actual and estimated speed in the case of ADRC control. One can clearly observe the robustness of this control as it rejects all disturbances.

10. Conclusion

The simulated prototype allows the variation of the mechanical speed of an induction machine without the use of a speed sensor. This prototype can be implemented in an electric vehicle. The combination of the 9-level converter, PWM, ADRC, PI, and Luenberger observer has helped reduce signal harmonics and electromagnetic torque ripples. The system has also been designed to reject the effects of internal and external disturbances. Operational principles and mathematical models (PV source, induction motor, ADRC control, PI control, indirect field-oriented control, and state observers) were analyzed in this study to ensure a good stability of the machine speed and a response time adapted to the evolution of this mechanical quantity. The results show that the system with ADRC is superior and more robust than the system with PI in various scenarios and disturbance conditions.

Conflict of interest

The authors declare no conflict of interest.

Author details

Abdellah Oukassi
Systems and Applications Engineering Laboratory (LISA), University Cadi Ayyad University, National School of Applied Sciences (ENSA), Marrakesh, Morocco

*Address all correspondence to: abdououkassi16@gmail.com

References

[1] Blaschke F. The principles of field orientation as applied to the new Tran vector closed loop control system for rotating field machines. Siemens Review. 1972;**34**:217-220

[2] Casadei D, Profumo F, Serra G, Tani A. FOC and DTC: Two viable schemes for induction motors torque control. In: Proceedings of the IEEE International Conference on Transactions and Power Electronics (TPW'02); 07 Nov 2022. IEEE; 2002. pp. 779-787

[3] Elamri O, Oukassi A, El Bahir L, El Idrissi Z. Combined vector and direct controls based on five-level inverter for high performance of IM drive. Journal of World Electric Vehicle. 2022;**13**(1):17. DOI: 10.3390/wevj13010017

[4] Oukassi A, El Bourhichi S, Eladnani M. Indirect vector control of induction motor fed by a five-level cascaded H-bridge inverter using the PV generators. In: Proceedings of the IEEE International Conference on Electrical and Information Technologies (ICEIT'20); Mois; Rabat. Morocco: IEEE; 2020. pp. 1-8

[5] Rodriguez J, Lai J-S, Peng F. Multilevel inverters. A survey of topologies, controls, applications. In: Proceedings of the IEEE International Conference on Trans. Ind. Electron (TIE'02); September 2002. IEEE; 2002. pp. 724-738

[6] Oukassi A, Elamri O, Boulghasoul Z. ADRC and IFOC control of a Sensorless induction motor driven by a multilevel converter using SVM approach and PV generators. Mathematical Problems. 2024;**2024**:15. DOI: 10.1155/2024/618332. [Article ID: 6183332]

[7] El Bourhichi S, Oukassi A, El Bahir L, El Adnani M. Active disturbance rejection control for a five-level cascaded H-bridge inverter fed induction motor Sensorless field-oriented. Mathematical Problems in Engineering. 2021;**2021**:13. DOI: 10.1155/2021/9925072. [Article ID: 9925072]

[8] Han J. From PID to active disturbance rejection control. In: Proceedings of the IEEE International Conference on Transactions on Industrial Electronics (TIE'09); 3, MARCH 2009. IEEE; 2009. pp. 900-906

[9] Saadatmand M, Mozafari B, Gharehpetian GB, Soleymani S. Optimal fractional-order PID controller of inverter-based power plants for power systems LFO damping. Turkish Journal of Electrical Engineering & Computer Sciences. 2020;**28**(36):485-499. DOI: 10.3906/elk-1907-31

[10] You J, Wu W, Wang Y. An adaptive Luenberger observer for speed-Sensorless estimation of induction machines. In: 2018 Annual American Control Conference (ACC'18); 27–29 June 2018; Milwaukee, WI, USA. 2018. pp. 2378-5861

[11] Fan X, Arcak M. Observer Design for Systems with multivariable monotone nonlinearities. Systems & Control Letters. 2003;**50**:319-330

[12] Sholapur S, Mohan KR, Narsimhegowda TR. Boost converter topology for pv system with perturb and observe mppt algorithm. IOSR journal of electrical and electronics. Engineering. 2014;**9**(4):50-56. DOI: 10.9790/1676-09425056

[13] Mollahasanoğlu M, Hakkı Mollahasanoğlu H, Okumuş İ. New

modified carrier-based level-shifted PWM control for NPC rectifiers considered for implementation in EV fast chargers. Turkish Journal Journal of Electrical Engineering & Computer Sciences. 2023;**31**:1255-1275. DOI: 10.55730/1300-0632.4046

[14] Sixing D, Wu B, Zargari NR, Cheng Z. A Fl-capacitor modular multilevel converter (FC-MMC) for medium-voltage motor drive. In: Proceedings of the IEEE International Conference Transactions on Power Electronics (TPW'17); January 2016. pp. 1-1

[15] Maheswari KT, Bharanikumar R, Arjun V, Amrish R, Bhuvanesh M. A comprehensive review on cascaded H-bridge multilevel inverter for medium voltage high power applications. Proceedings Journal. 2021;**45**(2): 2666-2670. DOI: 10.1016/j.mat pr.2020.11.519

[16] Deshmukh G, Chaturvedi P, Rajderkar V. Single phase cascaded H-bridge multilevel inverter topology. In: Proceedings of the IEEE International Conference on Futuristic Technologies (FT' 22); Belgaum; India; 25–27 November 2022. IEEE; 2022

[17] Podder S, Biswas MM, Rahman Khan MZ. A modified PWM technique to improve total harmonic distortion of multilevel inverter. In: Proceedings of the IEEE International Conference on Electrical and Computer Engineering (ECE'16); Dhaka, Bangladesh. ; December 2016. pp. 515-518

[18] Huang Y, Xue W. Active disturbance rejection control. Methodology and theoretical analysis. ISA Transactions. 2014;**53**:963-976. DOI: 10.1016/j. isatra.2014.03.003

[19] Xie J, Wei W, Liao P, Liu J. Design of ADRC controller for induction motor based on improved fal function. In: Proceedings of the IEEE International Conference o on Advanced Computational Intelligence (ACI'22), April 2022. pp. 216-220

[20] Fareh R, Khadraoui S, Abdallah MY, Baziyad M, Bettayeb M. Active disturbance rejection control for robotic systems: A review. Mechatronics. December 2021;**80**:102671. DOI: 10.1016/j.mechatronics.2021. 102671

[21] Sellami A, Zanzouri N. Diagnostic by luenberger observer using bond graph approach for a photovoltaic system. In: GECS International Conference on Green Energy Conversion Systems GECS'17; 12 October 2017

[22] Boulghasoul Z, Kandoussi Z, Elbacha A, Tajer A. Fuzzy improvement on Luenberger observer based induction motor parameters estimation for high performances Sensorless drive. Journal of Electrical Engineering & Technology. August 2020;**15**(2)

[23] Blaha P, Václavek P. Lyapunov-function-based flux and speed observer for AC induction motor sensorless control and parameters estimation. In: Proceedings of the IEEE International Conference on Transactions on Industrial Electronics (TEE'0) March 2006. pp. 138-145

Chapter 4

A Full-Order Terminal Sliding-Mode Observer-Based Optimal Resistance Estimation for High-Performance Control Applications

Saad Khadar, Mohaned Menzou and Tahar Abbassi

Abstract

This chapter introduces a sensorless field-oriented control (FOC) method for an induction motor with an open-end winding (IM-OEW). A full-order terminal sliding-mode (FOTSM) observer is proposed to accurately estimate the rotor flux, motor speed, and stator currents. The work presented in this chapter endeavors to elevate the sensorless control of the researched IM-OEW topology and fortify its ability to withstand parameter variations. Therefore, an estimation method has been developed to ascertain both the stator and rotor resistance during motor operation, thus contributing to the advancement of the sensorless FOC technique's exceptional performance and reducing system uncertainties. Combining the resistance estimation approach with the suggested control improves both speed estimates and control accuracy, especially at low speeds. The simulation results produced validate the improved performance and evident effectiveness of the developed control method in a range of various operating scenarios.

Keywords: FOC method, FOTSM observer, IM-OEW topology, sensorless control, parameter estimation

1. Introduction

Over the past two decades, researchers have focused on designing topologies and control strategies for multilevel inverters for various industrial applications [1–3]. Multilevel converter topologies are often utilized for high-power variable speed drives due to low dV/dt, and decreased harmonic distortion in output voltages [1–4]. One of the multilevel converter topologies is an open-end winding (OEW) structure. The OEW topology has sparked interest owing to its benefits over typical star or delta-connected alternatives. Advantages of employing dual inverters include multilevel output voltage, decreased switching frequency, and improved quality. This advantage reduces switching losses, eliminates common-mode voltages, and improves fault tolerance [5–9].

Several control strategies have been utilized for the IM-OEW, including field-oriented control [10], direct torque control [11], backstepping control [12], and sliding mode control [13]. The FOC approach is a simple and reliable solution that provides excellent dynamic performance by adjusting torque and flux independently across a wide range of speeds [11]. However, the FOC approach is sensitive to changes in motor resistances and load torque disturbances [2]. In addition, the FOC technique demands precise motor speed. The speed is often measured with a specific resolver on the motor shaft or calculated constantly using models or observers. In this sense, sensorless control minimizes the number of installed sensors, the extra electrical circuits, the connecting cables, and the mounting area [6]. Despite this, maintaining high-performance drives in sensorless control presents persistent issues, particularly at low speeds and with machine parameter variations.

Research on estimating motor speed, rotor resistance (R_s), and stator resistance (R_r) simultaneously has been conducted to enhance motor performance during resistance variations [2, 6, 14–17]. However, the proposed techniques established have considerable computing costs, longer execution times, are more complex, and poor performance during parameter variations, especially in low-speed operations.

This work aims to design and implement a robust sensorless control strategy using a FOTSM observer with motor resistance estimation. The proposed observers estimate motor speed, rotor flux, stator currents, stator resistance, and rotor resistance for OESW-TPIM under low-speed operation and parameter changes. The chapter is organized as follows: Section 2 presents the rotor-field-oriented IM-OEW model. Then Section 3 details the proposed observer to estimate the motor speed, the stator currents and the rotor flux. Section 4 describes the presented method for the motor resistance estimation. Section 5 introduces the simulation results obtained in this work. The conclusion of the presented work is given in the last section.

2. Rotor field-oriented IM-OEW model

The most often utilized control method in industrial drives, particularly for high-speed applications, is field-oriented control [10, 18–20]. The main objective of the FOC method for the IM-OEW topology is to manage the d-q components of stator currents in order to gain decoupled control of motor torque and rotor flux. The q-rotor flux becomes 0 if the rotor flux is aligned with the d-axis:

$$\begin{cases} \phi_{rd} = \phi_r \\ \phi_{rq} = 0 \end{cases} \tag{1}$$

The IM-OEW model in the d-q coordinate can be determined as follows:

$$\begin{cases} \dfrac{di_{sd}}{dt} = \mu \dfrac{\phi_r}{T_r} - \delta i_{sd} + \left(\omega + \dfrac{M i_{sq}}{\phi_r T_r}\right) i_{sq} + K V_{sd} \\ \dfrac{di_{sq}}{dt} = -\mu\omega\phi_r - \left(\omega + \dfrac{M i_{sq}}{\phi_r T_r}\right) i_{sd} - \delta i_{sq} + K V_{sq} \end{cases} \tag{2}$$

$$\begin{cases} \dfrac{d\omega}{dt} = \dfrac{n_p}{J}\left(\dfrac{n_p M \phi_r i_{sq}}{J L_r} - T_L\right) \\ \dfrac{d\phi_r}{dt} = -\dfrac{\phi_r}{T_r} + \dfrac{M i_{sd}}{T_r} \end{cases} \tag{3}$$

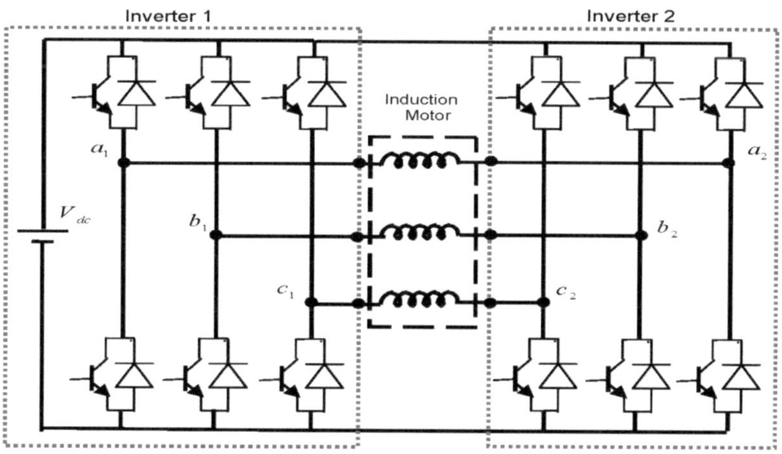

Figure 1.
The principal of the studied IM-OEW topology with one DC source.

where i_{sd} and i_{sq} are the stator currents in d-q coordinate; V_{sd} and V_{sq} are the stator voltages in d-q coordinate; ω is electric angular speed of the rotor; T_L is the load torque; n_p is the pole pairs number; M is the mutual inductance; ϕ_r is the rotor flux; R_s and L_s are the resistance and inductance of the stator; R_r, L_r, T_r are the resistance, inductance and time constant of the rotor. Also,

$$K = 1/\sigma L_s, \sigma = 1 - M^2/L_r L_s, \mu = K(M/L_r), \text{and } \delta = K\left(R_s + R_r M^2/L_r^2\right).$$

On the other side, the power circuit of the studied IM-OEW topology with only one DC source is shown in **Figure 1** [14]. It comprises a dual inverter and an IM, where the stator wire terminals (a_1, b_1, c_1) are linked to inverter 1, and (a_2, b_2, c_2) are coupled to inverter 2.

Thus, by using Kirchhoff's law on **Figure 1**, the phase voltages across each stator winding can be found as follows:

$$\begin{cases} V_{sa} = V_{a1} - V_{a2} \\ V_{sb} = V_{b1} - V_{b2} \\ V_{sc} = V_{c1} - V_{c2} \end{cases} \tag{4}$$

3. Developed FOTSM observer

The traditional sliding mode observer, while offering straightforward implementation and design flexibility [8, 21–23]. Unfortunately, the output injection in the conventional observer is a switching signal, which can generate large noise [24–27], a significant drawback that necessitates a solution. Therefore, this work proposes a full-order terminal sliding-mode observer that uses stator current and supplied voltage in α-β axes to estimate rotor flux, motor speed, and stator currents. The suggested improvement is not just a theoretical concept. It involves the practical application of the sensorless control, as demonstrated in **Figure 2**.

The IM-OEW model in α-β axes is determined as follows [24]:

$$\begin{cases} \dfrac{d\phi_{r\alpha\beta}}{dt} = -X + \dfrac{M}{T_r} i_{s\alpha\beta} \\ \dfrac{di_{s\alpha\beta}}{dt} = \mu X - \delta i_{s\alpha\beta} + K V_{s\alpha\beta} \end{cases} \tag{5}$$

The common matrix X contains the rotor flux, motor speed, and rotor time constant as:

$$X = \begin{pmatrix} 1/T_r & \omega \\ -\omega & 1/T_r \end{pmatrix} \begin{bmatrix} \phi_{r\alpha} \\ \phi_{r\beta} \end{bmatrix} \tag{6}$$

In this chapter, an FOTSM observer is created by getting the matrix X in order to estimate the stator currents and rotor flux as [28–32]:

$$\begin{cases} \dfrac{d\hat{\phi}_{r\alpha\beta}}{dt} = -\dfrac{1}{\mu} W_{\alpha\beta} + \dfrac{M}{T_r} \hat{i}_{s\alpha\beta} \\ \dfrac{d\hat{i}_{s\alpha\beta}}{dt} = W_{\alpha\beta} - \delta \hat{i}_{s\alpha\beta} + K V_{s\alpha\beta} \end{cases} \tag{7}$$

The switching signal used as the output injection in the traditional SM observer has the potential to introduce significant noise into the estimated M. In order to achieve smooth speed and rotor flux signals, this study suggests a chattering-free FOTSM observer that targets the matrix M. It also enhances the observer's dynamic performance and accuracy. An FOTSM observer is designed as follows [32]:

$$S = \frac{de_{s\alpha\beta}}{dt} + F_1 e_{s\alpha\beta} + F_2 \hat{i}_{s\alpha\beta}^{p/q} \tag{8}$$

where $e_{s\alpha\beta}$ is stator current error, $F_1 = \mathrm{diag}(f_{\alpha1}, f_{\beta1})$ and $F_2 = \mathrm{diag}(f_{\alpha2}, f_{\beta2})$ are employed to define the dynamics of sliding motion. Also, q and p are odds, and $0<q/p<1$.

The $W_{\alpha\beta}$ in Eq. (7) can be rewritten as follows:

$$W_{\alpha\beta} = \mu \begin{pmatrix} 1/T_r & \omega \\ -\omega & 1/T_r \end{pmatrix} \begin{bmatrix} \phi_{r\alpha} \\ \phi_{r\beta} \end{bmatrix} \tag{9}$$

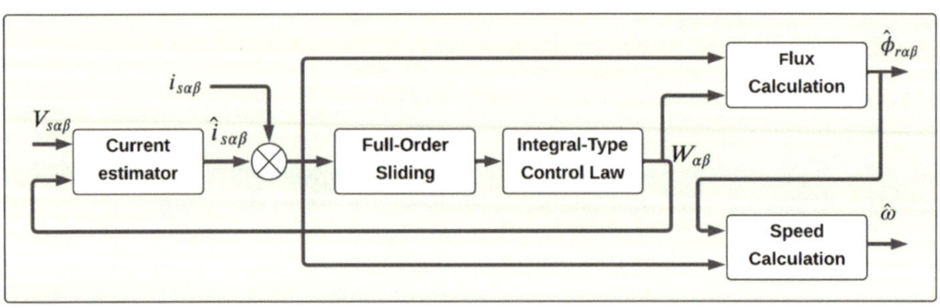

Figure 2.
Block diagram of developed FOTSM observer.

The required condition of the FOC technique is therefore provided by the following estimation of the rotor flux and motor speed as:

$$\begin{cases} \hat{\phi}_{r\alpha\beta} = \int_0^t \left(-\frac{1}{\mu} W_{\alpha\beta} + \frac{M}{T_r} \hat{i}_{s\alpha\beta} \right) \\ \omega_{est} = \left(W_\alpha \hat{\phi}_{r\beta} - W_\beta \hat{\phi}_{r\alpha} \right) / \left(\mu \| \phi_{r\alpha\beta} \| \right) \end{cases} \tag{10}$$

Finally, it is evident that a continuous output injection from observer $W_{\alpha\beta}$ can result in a smooth estimated speed.

4. Motor parameter estimation

Variations in motor characteristics, such as uncertainties in motor parameters, have a significant impact on the accuracy of the FOTSM observer and the performance of the FOC method [1, 2, 8]. The motor resistances, which are known to vary with temperature, can cause the stator resistance R_s and rotor resistance R_r to deviate from their initial values, as shown in Eq. (10). The precise value of R_s is crucial for accurately estimating stator flux components. Similarly, the rotor resistance can impact the FOC method [2], especially at low speeds [30–32]. In light of these challenges, this chapter proposes an approach to enhance the robustness and accuracy of the FOTSM observer and improve the performance of the sensorless FOC technique. **Figure 3** illustrates the basic concept of motor resistance estimation.

4.1 Stator resistance estimation

It is commonly known that changes in the temperature of the stator windings are the primary cause of variations in the motor stator resistance during motor operation [2, 6, 8, 14]. Consequently, in order to ensure that the estimation of stator resistance is performed online, a simple fuzzy logic algorithm is employed in this chapter. The following equation forms the error signals between the estimated stator currents and the measured stator currents [14]:

$$\begin{cases} z(n) = \hat{i}_s(n) - i_s(n) \\ \Delta z(n) = z(n) - z(n-1) \end{cases} \tag{11}$$

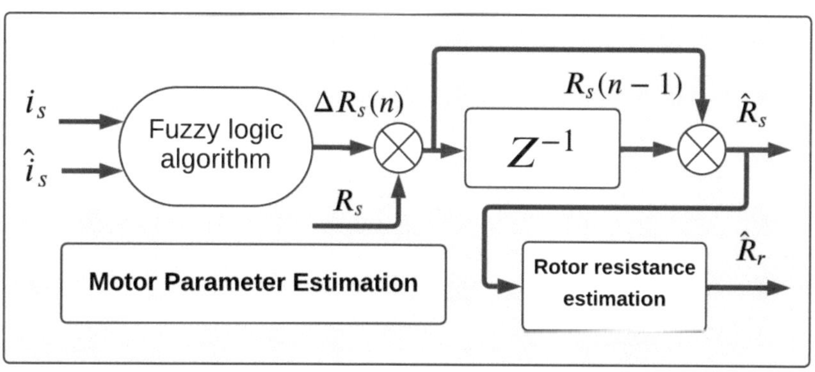

Figure 3.
Basic concept for motor resistance estimation.

The variables in the equation are defined as follows:

$$\begin{cases} i_s = \sqrt{(i_{s\alpha})^2 + (i_{s\beta})^2} \\ \hat{i}_s = \sqrt{(\hat{i}_{s\alpha})^2 + (\hat{i}_{s\beta})^2} \end{cases} \tag{12}$$

The stator resistance error is the fuzzy logic estimator's output and is defined as follows:

$$\Delta R_s(n) = R_s(n) - R_s(n-1) \tag{13}$$

1. Fuzzification: the estimator has two inputs: z(n), Δz(n), and for the fuzzified output R_s(n). Five fuzzy language variables (NB, NS, ZO, PS, PB), define the language variable of z(n), Δz(n), and ΔR_s(n), as shown in **Figure 4**.

2. Interfacing: the static and dynamic behavior of the system to be regulated determines which inference method should be used. Therefore, the Max-Min fuzzy inference approach is used for the R_s variation due to its ease of implementation.

3. Defuzzification: our selection is based on the balanced heights center of attraction method, which is the most widely used defuzzification technique. Its simplicity and low computational overhead make it our first pick.

4.2 Rotor resistance estimation

In this study, a simple and effective approach is presented for real-time rotor resistance identification, where the objective is to reduce the cost of the algorithm used without additional sensors or signals. The temperature coefficients of the metals employed in the stator and rotor windings, respectively, at 20°C are λ_1 and λ_2. If the rotor conductor or bars are composed of aluminum, then $\lambda_1 = 0.00429$ and $\lambda_2 = 0.00386$. If both windings are constructed of copper, then $\lambda_1 = \lambda_2 = 0.00386$. Thus, the rotor resistance value can be determined continuously to account for any changes due to temperature as follows [6]:

$$\hat{R}_r = R_r\left(1 + \frac{\lambda_2}{\lambda_1}\left(\frac{\hat{R}_s}{R_s} - 1\right)\right) \tag{14}$$

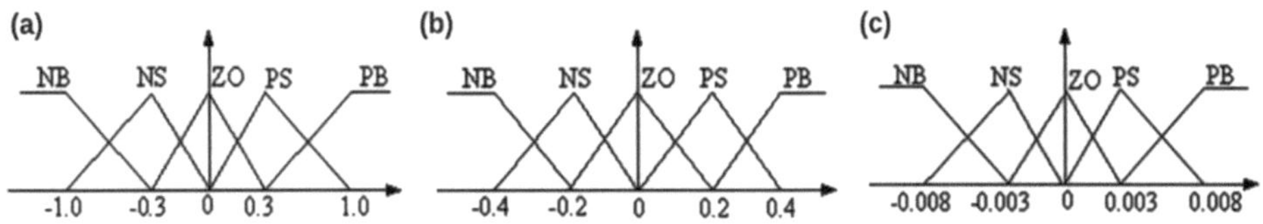

Figure 4.
The membership functions.

5. Simulation results

Figure 5 displays the general schematic representation of the suggested sensorless control for the investigated IM-OEW structure, which is based on the control-observer theory discussed earlier.

5.1 Load torque change

In order to demonstrate the effectiveness and the correctness of proposed sensorless control under load torque change, a MATLAB/Simulink simulation model was developed to model the whole control system shown in **Figure 5**. The used speed reference is shown in **Figure 6(a)** which is a step fixed at 120 rad/s and following the application of a load torque at t = 2 s equal to 2 N.m **Figure 6(b)** illustrates the measured and estimated speeds. It can be remarked that the estimated value of motor speed quickly tracks the measured value of motor speed, where it can be observed that the speed estimation error is around zero during steady state, as shown in **Figure 6(c)**. Hence the FOTSM observer given is effective. One can notice that the motor torque follows the load torque very well as shown in **Figure 6(d)** with small overshoots.

5.2 Motor parameter variations

In order to prove the robustness of the proposed control and estimation systems to parameter variations, the studied IM-OEW topology is started with its nominal value of stator resistance, after t = 2 s, stator resistance is changed to half its nominal value. Similarly, the rotor resistance is set suddenly at 1.5 × R_{sn} at t = 2 s. In this sense, the motor resistances used in the proposed controller design are 50% smaller than the nominal values. It is maintained that the reference rotor-flux and speed are 1.2 Wb and 10 rad/s, respectively.

Figure 7(b) shows the measured and estimated speeds, it can be seen that the good ability to maintain the estimated speed close to the measured one during low-speed

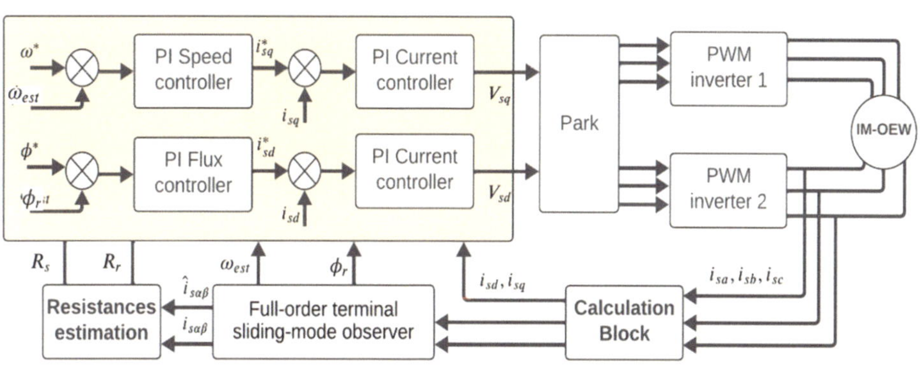

Figure 5.
Block diagram of proposed sensorless control for the IM-OEW topology.

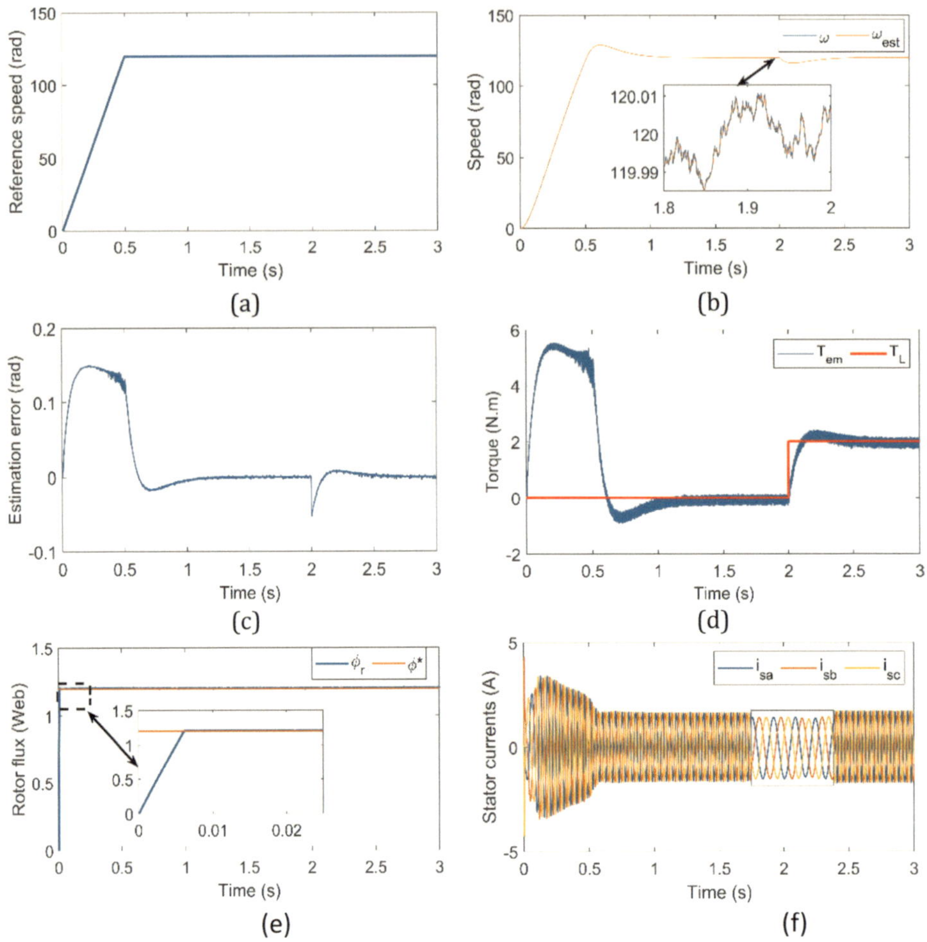

Figure 6.
Simulation results of sensorless control based on FOTSM observer under load torque change.

operation. Furthermore, results indicate that due to the accurate motor speed estimation, the IM-OEW drive does not lose stability during operation at low speeds. This proves the robustness and effectiveness of the proposed control law under parameter variations, which can change from the nominal ones according to motor temperature and operating conditions. **Figure 7(c)** confirms that the fuzzy stator resistance estimator is able to estimate the changes in stator resistance. Furthermore, the rotor resistance R_r can be identified within a short interval, as shown in **Figure 7 (d)**. The estimated speed is compared with the measured one as presented in **Figure 7(e)**. Clearly, the speed estimation error is found to be less than 0.014 rad/s in the steady-state. **Figure 7(f)** shows the load torque and motor torque of the studied IM-OEW topology. **Figure 7(g)** illustrates the reference and estimated rotor flux. It can be observed that even if the motor resistance changes, the proposed method still gives a good estimate of rotor flux. As can be seen in **Figure 7(h)**, the stator current falls from 1.2 amps to about 1.34 amps due to the changes in the rotor and stator resistance.

DOI: http://dx.doi.org/10.5772/intechopen.1006065

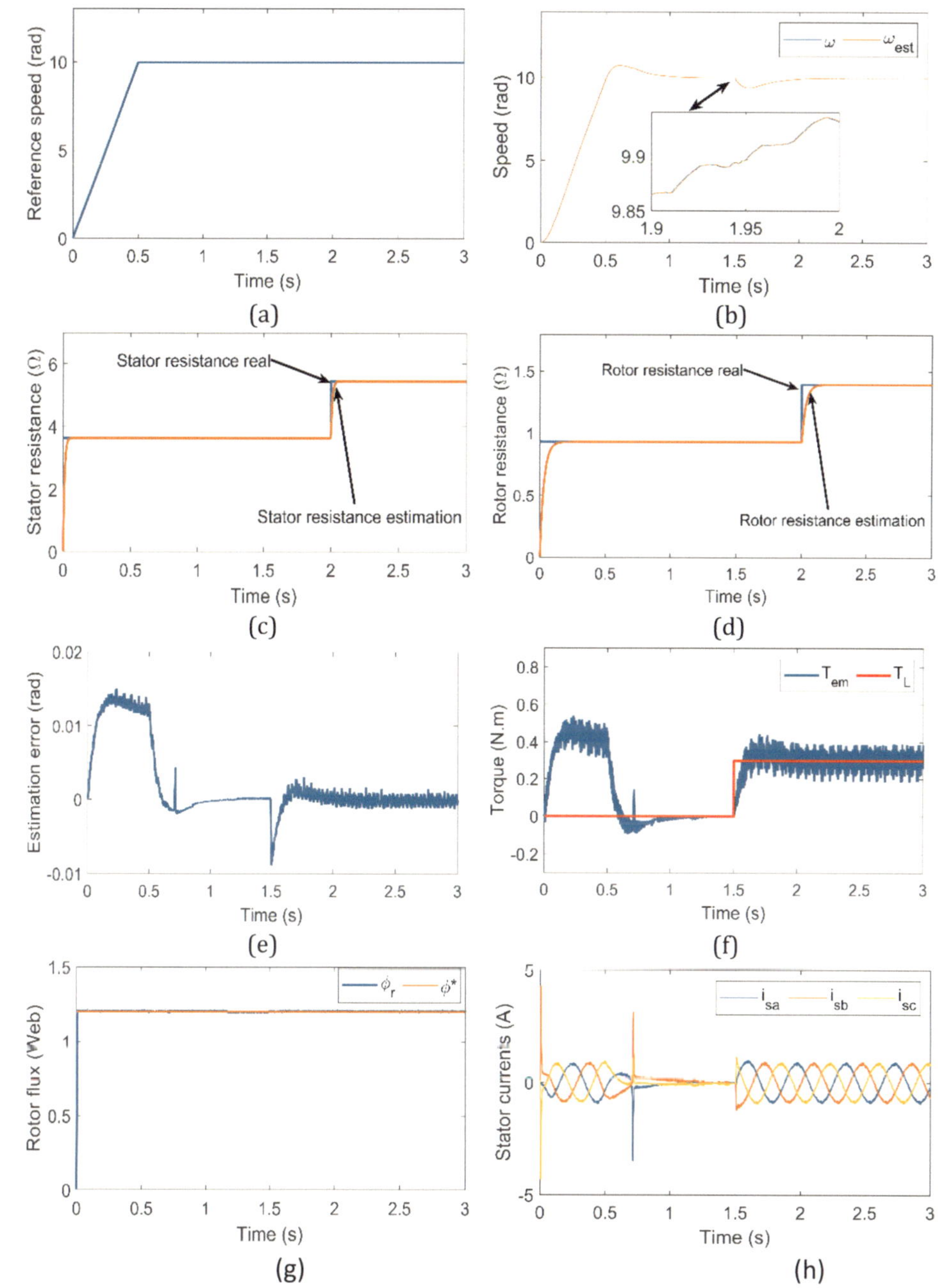

Figure 7.
Simulation results of sensorless control based on FOTSM observer under IM-OEW parameter variations.

6. Conclusions

In this chapter, an improved full-order terminal sliding-mode observer was presented for the sensorless FOC control of the IM-OEW topology. The purpose of the proposed observer was to simultaneously estimate the motor speed, rotor flux and stator currents as well as track the stator and rotor resistance variations in real-time. The obtained simulation results demonstrate that the sensorless FOC method with the developed resistance estimation approach offers excellent results in terms of resistance variations and speed tracking with the lowest estimation error compared to the previous works. Therefore, it can be ensured that the approach developed in this chapter will be both feasible and beneficial in real-world engineering applications.

Author details

Saad Khadar[1*], Mohaned Menzou[2] and Tahar Abbassi[1]

1 Applied Automation and Industrial Diagnostics Laboratory (LAADI), Ziane Achour University, Djelfa, Algeria

2 Faculty of Applied Sciences, Department of Computer Science, WSB University, Dąbrowa Górnicza, Poland

*Address all correspondence to: saad.khadar@univ-djelfa.dz

References

[1] Khadar S. Influence of a different fault scenarios on the properties of multi-phase induction machine. Algerian Journal of Engineering and Technology. 2020;**2**:11-21. DOI: 10.5281/zenodo.3647802

[2] Dehghan-Azad E, Gadoue S, Atkinson D, Slater H, Barrass P, Blaabjerg F. Sensorless control of IM based on stator-voltage MRAS for limp-home EV applications. IEEE Transactions on Power Electronics. 2018;**33**:1911-1921

[3] Khadar S, Kouzou A, Benguesmia H. Fuzzy stator resistance estimator of induction motor fed by a three levels NPC inverter controlled by direct torque control. In: Proceedings of the IEEE International Conference on Applied Smart Systems (ICASS); 24–25 November 2018. Medea, Algeria: IEEE; 2018

[4] Mohamed IDA, Ahmed E, Ahmed MM, Radu B, Shehab A. Zero−/low-speed operation of multiphase drive systems with modular multilevel converters. IEEE Access. 2019;**7**:14353-14365

[5] Menzou M, Khadar S, Mohammed Elamine K. RT-lab based real-time simulation of a new solar PV Array powered open-end winding induction motor using a boost DC-DC converter. In: Proceedings of the IEEE 1st International Conference on Renewable Solutions for Ecosystems: Towards a Sustainable Energy Transition (ICRSEtoSET); 06–08 May 2023. Djelfa, Algeria: IEEE; 2023

[6] Listwan J, Pieńkowski K. Field-oriented control of five-phase induction motor with open-end stator winding. Archives of Electrical Engineering. 2016; **65**(03):395-410Te

[7] Siddique A, Yadava G, Singh B. Applications of artificial intelligence techniques for induction machine stator fault diagnostics: Review. IEEE Transactions on Energy Conversion. 2005;**20**(4):759-767

[8] Khadar S, Kouzou A, Rezaoui M, Hafaifa A. Fault-tolerant sensorless sliding mode control by parameters estimation of an open-end winding five-phase induction motor. Modeling, Measurement, and Control A. 2019;**92** (2–4):6-15

[9] Jingchuan L, Longya X, Zhang Z. An adaptive sliding mode observer for induction motor sensorless speed control. IEEE Transactions on Industry Applications. 2005;**41**:1039-1046

[10] Kim YR, Sul SK, Park MH. Speed sensorless vector control of an induction motor using an extended Kalman filter. IEEE Transactions on Industry Applications. 1994;**30**:1225-1233

[11] Khadar S, Kouzou A, Hafaifa A. Sensorless direct torque control of induction motor with an open-end stator winding using an adaptive Luenberger observer. In: Proceedings of the IEEE 15th International Multi-Conference on Systems, Signals & Devices (SSD). Yasmine Hammamet, Tunisia; 2018. pp. 1355-1362. DOI: 10.1109/SSD.2018.8570691

[12] Khadar S, Kouzou A. Fault-tolerant control of asynchronous machine taking into account faults. In: Proceedings of the IEEE International Conference on Electrical Sciences and Technologies in Maghreb (CISTEM). Algiers, Algeria: IEEE; 2018. pp. 28-31. DOI: 10.1109/CISTEM.2018.8613442

[13] Khadar S, Kobbi M, Menzou M, Senni A, Zakaria S, Ahmed H. Mover position control of open-end stator winding integrated to wind power system using dSPACE 1103 control board. In: Proceedings of the IEEE 1st International Conference on Renewable Solutions for Ecosystems: Towards a Sustainable Energy Transition. Djelfa, Algeria; 2023. pp. 1-5. DOI: 10.1109/ICRSEtoSET56772.2023.10525329

[14] Mohamed SZ, Mahmoud MK, Shokry SS, Hussain AY. Wide-speed-range estimation with online parameter identification schemes of sensorless induction motor drives. IEEE Transactions on Industrial Electronics. 2009;**56**:1699-1707

[15] Lascu C, Boldea I, Blaabjerg F. Direct torque control of sensorless induction motor drives: A sliding-mode approach. IEEE Transactions on Industry Applications. 2004;**40**:582-590

[16] Hamdi E, Ramzi T, Atif I, Mohamed M. F: Real time implementation of indirect rotor flux-oriented control of a five-phase induction motor with novel rotor resistance adaption using sliding mode observer. Journal of the Franklin Institute. 2018;**355**:2112-2141

[17] Bo F, Zhumu F, Leipo L, Jiangtao F. The full-order state observer speed-sensorless vector control based on parameters identification for induction motor. Measurement and Control. 2019; **52**(3/4):202-211

[18] Khadar S, Kaddouri AM, Kouzou A, Hafaifa A, Kennel R, Abdelrahem M. Experimental validation of different control techniques applied to a five-phase open-end winding induction motor. Energies. 2023;**16**:5288. DOI: 10.3390/en16145288

[19] Bennassar A, Abbou A, Akherraz M, Barara M. Sensorless backstepping control using an adaptive Luenberger observer with three levels NPC inverter. International Journal of Computer, Electrical, Automation, Control and Information Engineering. 2013;7:1171-1177

[20] Khadar S, Abdelaziz A, Elbarbary ZMS, Mossa MA. An improved sensorless nonlinear control based on SC-MRAS estimator of open-end winding five-phase induction motor fed by dual NPC inverter: Hardware-in-the-loop implementation. Machines. 2023;**11**(4):469

[21] Khadar S, Kouzou A, Refaat SS, Abu-Rub H. Real-time digital simulation for Sensorless control scheme based on reduced-order sliding mode observer for dual star induction motor. In: Proceedings of the IEEE 46th Annual Conference of the IEEE Industrial Electronics Society; Singapore. 2020. pp. 4869-4874. DOI: 10.1109/IECON43393.2020.9254945

[22] Zhonggang Y, Guoyin L, Yanqing Z, Jing L, Xiangdong S, Yanru Z. A speed and flux observer of induction motor based on extended Kalman filter and Markov chain. IEEE Transactions on Power Electronics. 2017;**32**:7096-7117

[23] Mihai Comanescu C. An induction-motor speed estimator based on integral sliding-mode current control. IEEE Transactions on Industrial Electronics. 2009;**56**:3414-3423

[24] Mohamed M, Ahmed A, Hassan M. MRAS-based sensorless speed backstepping control for induction machine, using a flux sliding mode observer. Turkish Journal of Electrical Engineering and Computer Sciences. 2015;**23**(1):187-200

[25] Mitronikas E.D, Safacas AN, Tatakis EC. A new stator resistance tuning method for stator-flux-oriented vector-controlled induction motor drive. IEEE Transactions on Industrial Electronics. 2001;**48**:1148-1157

[26] Vasic V, Vukosavic SN, Levi E. A stator resistance estimation scheme for speed sensorless rotor flux oriented induction motor drives. IEEE Transactions on Energy Conversion. 2003;**18**:476-483

[27] Mir S, Elbuluk ME, Zinger DS. PI and fuzzy estimators for tuning the stator resistance in direct torque control of induction machines. IEEE Transactions on Power Electronics. 1998; **13**:279-287

[28] Guidi G, Umida H. A novel stator resistance estimation method for speed-sensorless induction motor drives. IEEE Transactions on Industry Applications. 2000;**36**:1619-1627

[29] Xepapas S, Kaletsanos A, Xepapas F, Manias S. Sliding-mode observer for speed-sensorless induction motor drives. IEEE Proceedings- Control Theory and Applications. 2003;**150**:611-617

[30] Barut M, Demir R, Zerdali E, Inan R. Real-time implementation of bi input-extended Kalman filter-based estimator for speed-sensorless control of induction motors. IEEE Transactions on Industrial Electronics. 2012;**59**:4197-4206

[31] Kubota H, Matsuse K. Speed sensorless field-oriented control of induction machines using flux observer. In: IEEE Ind. Elect., Contand Instr, IECON '94; Bologna, Italy. 2002

[32] Lascu C, Boldea I, Blaabjerg F. Direct torque control of sensorless induction motor drives: A sliding-mode approach. IEEE Transactions on Industry Applications. 2004;**40**:582-590

Chapter 5

Predictive Control for Induction Machines

Marco Rivera, Emrah Zerdali, Patrick Wheeler and José Rodríguez

Abstract

In the last decades, the application of fast modern microcontrollers has been continuously growing, allowing the development and implementation of new and more intelligent control strategies as an alternative to conventional techniques for electrical machines and power converters. Model predictive control is one of these powerful and attractive alternatives that has received much attention in recent years. The use of predictive control offers several interesting advantages: it is an intuitive control approach, it does not need linear controllers and modulators, and it is possible to easily include nonlinearities and restrictions in the control law. It is expected that the advantages of predictive control will lead to industrial applications very shortly. In this chapter, new advances and trends in the application of model predictive control for induction machines will be presented.

Keywords: direct model predictive control, induction machine, modulated model predictive control, sensorless control, weighting factor elimination

1. Introduction

Model predictive control (MPC) uses the mathematical model of the system to predict the future performance of the controlled variables in a finite time horizon. This control strategy has the ability to optimize the control performance by considering constraints, minimizing a cost function, and providing fast dynamic responses. These characteristics made MPC particularly suitable for renewable energy systems, power converters, and electrical drives such as induction machine drive systems.

Recent advances and developments in MPC for induction machines include enhanced predictive control schemes such as improved predictive torque and flux control strategies for electric vehicles or industrial automation applications [1–3]. There are recent publications focused on the improvement of the MPC strategy by implementing long-horizon prediction with the k-best-sphere decoding algorithm (kSDA) to reduce the computational burden by transforming the optimization problem into an integer least-squares (ILSs) problem [4]. Recent works by Xu et al. [5] and Aboelhassan et al. [6] have been presented to improve the performance of the system by reducing high current harmonics reduce the ripple by including sensibility to the magnetizing inductance parameter. In a study by Aboelhassan et al. [6], modulated model predictive control (M2PC) to improve the dynamic performance by including discontinuous pulse with modulation (DPWM) for low-inductance electric machines.

In order to improve the performance of the system, the prediction model, and the dynamic response, Hamad et al. [7] have recently proposed a full-order sliding mode control (FOSMC)-based finite set model predictive flux control (FS-MFPC). The MPC technique has been improved by proposing several fault diagnosis and tolerant techniques [8–11]. A recent idea has proposed a free-model predictive control and compensation technique to eliminate the necessity of a mathematical model of the system for the prediction model, making the strategy robust under parameter mismatches in the prediction model [12, 13].

AI-based techniques, like artificial intelligence and machine learning, have been integrated into MPC to predict disturbances and adjust control parameters dynamically, enhancing the adaptability of the control strategy in variable conditions [14, 15].

In this chapter, a review of different predictive control techniques for induction machines is presented, highlighting the control scheme as well as the performance in both the dynamic and steady states.

2. Induction machine model fed by a voltage source inverter

This section introduces the mathematical model of an induction machine fed by a two-level voltage source inverter (VSI). For this purpose, the mathematical model of an induction machine is first presented, followed by the VSI model.

2.1 Induction machine model

An induction machine can be mathematically represented by a model based on the following general form:

$$\begin{aligned} \dot{\boldsymbol{x}}_t &= \boldsymbol{f}(\boldsymbol{x}_t, \boldsymbol{u}_t) + \boldsymbol{w}_t \\ \boldsymbol{z}_t &= \boldsymbol{h}(\boldsymbol{x}_t) + \boldsymbol{v}_t \end{aligned} \tag{1}$$

$\boldsymbol{f}$ is the nonlinear transition matrix function, $\boldsymbol{h}$ is the measurement matrix function, $\boldsymbol{x}_t$ is the state vector, $\boldsymbol{u}_t$ is the control input vector, $\boldsymbol{w}_t$ and $\boldsymbol{v}_t$ are the system and measurement noises, respectively.

When this model is defined in the stator stationary reference ($\alpha\beta$) frame, the vectors are as follows:

$$\boldsymbol{x}_t = \begin{bmatrix} i_{s\alpha} & i_{s\beta} & \omega_m & \psi_{r\alpha} & \psi_{r\beta} \end{bmatrix}^T,$$
$$\boldsymbol{u}_t = \begin{bmatrix} v_{s\alpha} & v_{s\beta} \end{bmatrix}^T, \boldsymbol{h}(\boldsymbol{x}_t) = \begin{bmatrix} i_{s\alpha} & i_{s\beta} & \omega_m \end{bmatrix}^T,$$

and

$$
f(x_t, u_t) = \begin{bmatrix} -\frac{1}{T_\sigma} i_{s\alpha} + \frac{k_r}{T_r L_\sigma} \psi_{r\alpha} + \frac{k_r}{L_\sigma} \omega_r \psi_{r\beta} + \frac{1}{L_\sigma} v_{s\alpha} \\ -\frac{1}{T_\sigma} i_{s\beta} + \frac{k_r}{L_\sigma T_r} \psi_{r\beta} - \frac{k_r}{L_\sigma} \omega_r \psi_{r\alpha} + \frac{1}{L_\sigma} v_{s\beta} \\ \frac{3}{2} k_r \frac{p_p}{J_t} \left(\psi_{r\alpha} i_{s\beta} - \psi_{r\beta} i_{s\alpha} \right) - \frac{B_t}{J_t} \omega_m - \frac{1}{J_t} \tau_l \\ \frac{L_m}{T_r} i_{s\alpha} - \frac{1}{T_r} \psi_{r\alpha} - \omega_r \psi_{r\beta} \\ \frac{L_m}{T_r} i_{s\beta} - \frac{1}{T_r} \psi_{r\beta} + \omega_r \psi_{r\alpha} \end{bmatrix},
$$

$v_{s\alpha}$ and $v_{s\beta}$ are the stator voltages, $i_{s\alpha}$ and $i_{s\beta}$ are the stator currents, $\psi_{r\alpha}$ and $\psi_{r\beta}$ are the rotor fluxes, ω_m and ω_r $(= p_p \omega_m)$ are the mechanical and electrical angular rotor speeds, respectively. τ_l is the load torque, T_r $(= L_r/R_r)$ is the rotor time constant, T_σ $(= L_\sigma/R_\sigma)$ is the transient time constant of the stator, and k_r $(= L_m/L_r)$ is the rotor coupling factor. R_σ $(= R_s + k_r^2 R_r)$ is the equivalent resistance referred to the stator side, L_σ $(= L_s - k_r L_m)$ is the leakage inductance, R_s and R_r are the stator and rotor resistances, respectively. $L_m, L_s,$ and L_r are the mutual, stator, and rotor inductances, respectively, p_p is the pole-pairs, J_t and B_t are the total inertia and viscous friction of both load and motor, respectively.

Since the above model is used for estimations and predictions, it needs to be discretized. For this purpose, the Euler approximation given below is utilized in this chapter as indicated here:

$$
\dot{x}_t \approx \frac{x_{k+1} - x_k}{T} \tag{2}
$$

where T is the sampling time. Eventually, the resulting discrete model is as follows:

$$
x_{k+1} = x_k + T \cdot f(x_t, u_t) \tag{3}
$$

2.2 VSI model

A two-level VSI, as depicted in **Figure 1a**, can be used to supply the machine. With this topology, seven distinct voltage vectors can be produced, as illustrated in **Figure 1b**, with a total of eight possible switching combinations. Among these vectors, v_0 and v_7 are passive vectors, while the others are active vectors. The inverter output voltage, v_s, can be determined using the switching combination $S_x \in \{S_a, S_b, S_c\}$ and the dc-link voltage V_{dc} as follows:

$$
v_s = \frac{2}{3} V_{dc} \left(S_a + a S_b + a^2 S_c \right) \tag{4}
$$

where a denotes the phase shift of 120 electrical degrees.

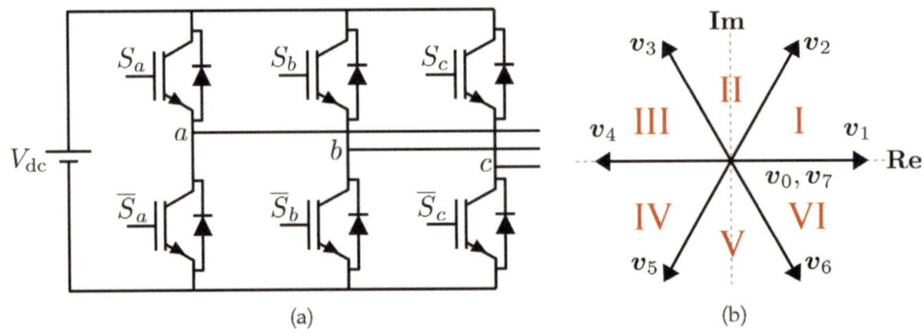

Figure 1.
Two-level voltage source inverter: (a) circuit topology, (b) possible VVs.

3. Classic MPC schemes

Various finite control set MPC (FCS-MPC) schemes have been proposed to overcome the disadvantages of previous MPC schemes. It is known that model predictive current control (MPCC) and model predictive torque control (MPTC) are among the earliest schemes used in the control of power converters and electric drives. Model predictive direct speed control (MPDSC) advances these schemes by eliminating the cascaded control structure and enhancing the disturbance rejection capability of the electric drive system. Therefore, this section introduces these three MPC schemes as classic MPC approaches.

3.1 MPCC scheme

The MPCC scheme is undoubtedly the most common FCS-MPC scheme suitable for power converters and electrical drives. A representation of the MPCC scheme for the 2L-VSI feeding an induction machine is presented in **Figure 2**. The MPCC is simple and requires lower computational complexity in comparison to other MPC schemes for electrical drives.

This control strategy uses the stator current predictions to evaluate the possible VVs. For this purpose, the stator current vector is predicted as follows:

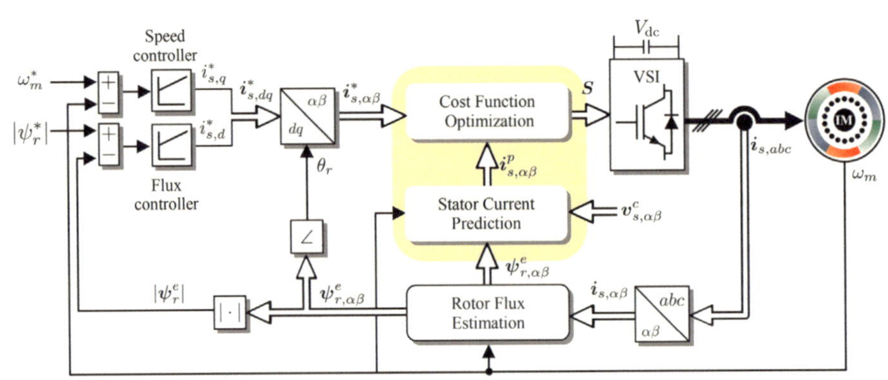

Figure 2.
Block diagram of the MPCC scheme.

$$\boldsymbol{i}^p_{s,k+1} = \left(1 + \frac{T}{T_\sigma}\right)\boldsymbol{i}_{s,k} + \frac{T}{T_\sigma}\left(\frac{1}{R_\sigma}\left(\left(\frac{k_r}{T_r} - jk_r\omega_r\right)\boldsymbol{\psi}^e_{r,k} + \boldsymbol{v}_{s,k}\right)\right) \tag{5}$$

It is clear that rotor flux $\boldsymbol{\psi}_r$ is needed to predict stator current $\boldsymbol{i}^p_s$. For this purpose, using the following current model of induction machine is a well-accepted way to estimate the rotor flux.

$$\boldsymbol{\psi}^e_{r,k+1} = \boldsymbol{\psi}^e_{r,k} + T\left[\frac{L_m}{T_r}\boldsymbol{i}_{s,k} - \left(\frac{1}{T_r} - j\omega_{r,k}\right)\boldsymbol{\psi}^e_{r,k}\right] \tag{6}$$

The following cost function consisting of only stator current errors is used to evaluate the performance of each VV.

$$g = \left(i^*_{s\alpha} - i^p_{s\alpha}\right)^2 + \left(i^*_{s\beta} - i^p_{s\beta}\right)^2 + i_{oc,k+1} \tag{7}$$

$i^*_{s\alpha}$ and $i^*_{s\beta}$ represent the reference trajectories of the stator currents, which are obtained by applying the inverse Park transformation to the outputs of the speed and flux controllers. The basic cost function only includes current error terms and does not require any weighting factors (WFs) since both terms are of the same unit.

However, the flexible structure characteristic of MPC can be expanded by additional control objectives (ACOs) in different units, such as overcurrent protection term, switching frequency regulation term, input voltage balancing term, and so on. In this case, a control objective i_{oc}, defined mathematically in Eq. (8), is included in Eq. (7) to prevent excessive currents in the induction machine.

$$i_{oc,k+1} = \begin{cases} 0, & \text{if } |\boldsymbol{i}^p_{s,k+1}| \leq i_{s,\max} \\ \infty, & \text{if } |\boldsymbol{i}^p_{s,k+1}| > i_{s,\max} \end{cases} \tag{8}$$

where $i_{s,\max}$ represents the maximum permissible current limit.

The system requires an outer controller to generate the current references as depicted in **Figure 2**. A linear proportional-integral (PI) controller is used for the control of the mechanical speed and another for the control of the rotor flux.

The control performance of the MPCC scheme is presented in **Figure 3** with a sampling time of $T = 25\ \mu s$. The test considers the magnetization of the machine, an initial speed step to $w_m = 100 rad/s$ at $t = 0.05s$ followed by a speed reversal at $t = 0.6s$. A load torque is applied at $t = 0.2s$ and another at $t = 0.4s$. Here is observed a very good performance of the currents to their respective references with sinusoidal waveforms almost all the time and fast dynamic response. A good tracking of the speed and stator flux to their respective references is also observed.

3.2 MPTC scheme

The MPTC scheme is another popular FCS-MPC strategy with high dynamic response and low torque ripple for the control of electrical drives. A representation of the MPTC for the 2L-VSI feeding an induction machine is presented in **Figure 4**. This control strategy uses the same steps used in predicting the stator current in the MPCC scheme. But, unlike MPCC, which controls the current in the internal control loop, MPTC directly controls electromagnetic torque and flux.

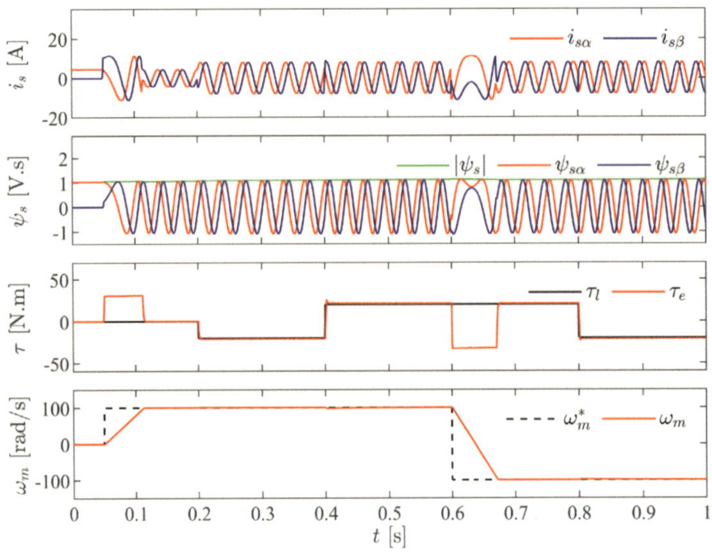

Figure 3.
Control performance of the MPCC scheme ($T = 25\ \mu s$).

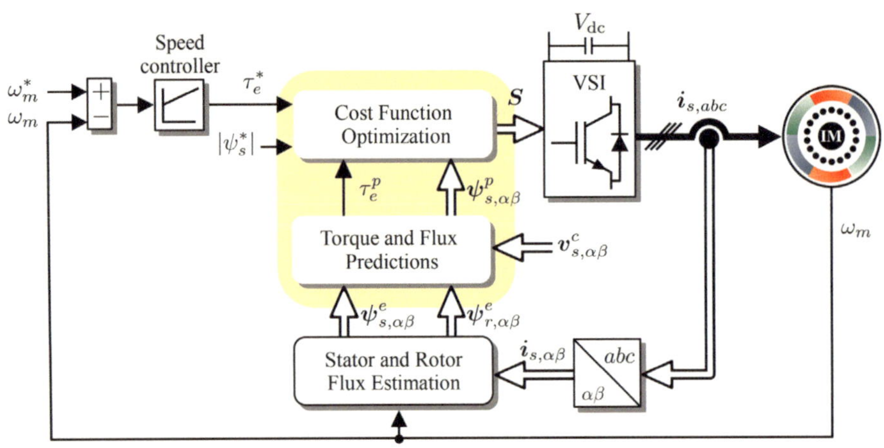

Figure 4.
Block diagram of the MPTC scheme.

The stator flux $\boldsymbol{\psi}_s$ must be predicted to perform flux control. This is carried out as follows:

$$\boldsymbol{\psi}_{s,k+1}^p = \boldsymbol{\psi}_{s,k}^e + T\boldsymbol{v}_{s,k} - TR_s\boldsymbol{i}_{s,k} \tag{9}$$

where

$$\boldsymbol{\psi}_{s,k}^e = k_r\boldsymbol{\psi}_{r,k}^e + L_\sigma\boldsymbol{i}_{s,k} \tag{10}$$

The electromagnetic torque τ_e can now be calculated using $\boldsymbol{i}_{s,k+1}^p$ and $\boldsymbol{\psi}_{s,k+1}^p$ as follows:

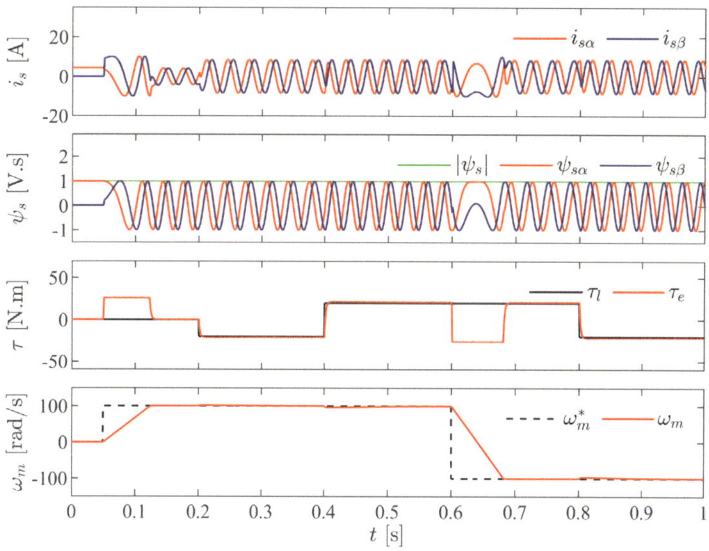

Figure 5.
Control performance of MPTC scheme ($T = 25\ \mu s$).

$$\tau^p_{e,k+1} = \frac{3}{2}p_p\, m\left\{\left(\boldsymbol{i}^p_{s,k+1}\right)\mathrm{conj}\left(\boldsymbol{\psi}^p_{s,k+1}\right)\right\} \tag{11}$$

Finally, the following cost function is used to evaluate the performance of each VV.

$$g = \left|\tau^*_e - \tau^p_{e,k+1}\right| + \lambda_\psi\left|\boldsymbol{\psi}^*_s| - |\boldsymbol{\psi}^p_{s,k+1}|\right| + i_{oc,k+1} \tag{12}$$

τ^*_e and $\boldsymbol{\psi}^*_s$ are the reference trajectories for electromagnetic torque and stator flux, respectively. λ_ψ is the weighting factor associated with the stator flux error.

The control performance of the MPTC scheme is presented in **Figure 5** by considering the same operational conditions as the previous case. A good tracking of both the stator flux and electrical torque to their respective references is observed all the time with a very fast dynamic response under step changes, which can be explained because the predictive controller does not require any modulator. The speed is handled by using an external linear PI controller, observing also good performance in the four quadrants.

3.3 MPDSC scheme

The MPDSC scheme needs load torque information to predict mechanical speed. Like previous classic MPC schemes, it also requires rotor flux information. To address this, an extended Kalman filter (EKF) observer is designed to estimate both rotor flux and load torque and is integrated into the MPDSC scheme. In the following sections, first the MPDSC scheme is presented, followed by the EKF observer.

3.3.1 MPDSC scheme

A representation of the model predictive direct speed control scheme (MPDSC) for the 2L-VSI feeding an induction machine is presented in **Figure 6**. This control strategy follows similar steps to the MPTC scheme for predicting stator current, stator flux, and

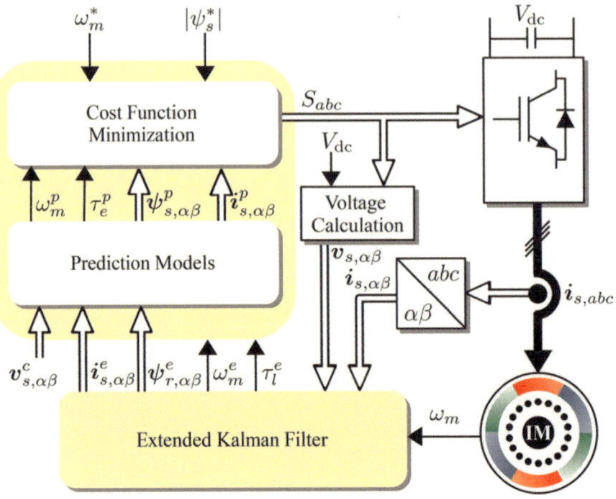

Figure 6.
Block diagram of the MPDSC scheme.

electromagnetic torque. However, unlike the MPTC scheme, MPDSC eliminates the outer control loop and provides a non-cascade form, providing better dynamic performance and the ability to reject disturbances [16]. The MPDSC scheme utilizes the rotor flux estimated by an EKF observer, which is introduced in the subsequent section. For direct speed control, speed prediction is also necessary in the MPDSC scheme. To achieve this, the equation of motion defined in the discrete form below, can be used.

$$\omega^p_{m,k+1} = \omega^e_{m,k} + \frac{T}{J_t}\left(\tau^p_{e,k+1} - \tau^e_{l,k+1}\right) \tag{13}$$

The prediction models in the MPDSC scheme utilize filtered stator current $\left(\boldsymbol{i}^e_s\right)$ and mechanical speed $\left(\omega^e_m\right)$ instead of the measured values. This approach aims to suppress noise in the predictions, leading to more accurate control.

Finally, $\omega^p_{m,k+1}$, $\boldsymbol{\psi}^p_{s,k+1}$ and $\tau^p_{e,k+1}$ are substituted into the following cost function to determine the optimal switching combination for the next sampling interval. The objective is to minimize errors in speed, flux, and torque simultaneously.

$$g = \lambda_\omega\left(\omega^*_m - \omega^p_{m,k+1}\right)^2 + \lambda_\psi\left(|\boldsymbol{\psi}^*_s| - |\boldsymbol{\psi}^p_{s,k+1}|\right)^2 + \lambda_\tau\left(\tau^p_{e,k+1} - \tau^e_{l,k+1}\right)^2 + i_{oc,k+1} \tag{14}$$

where ω^*_m is the reference speed trajectory, and λ_ω, λ_ψ, and λ_τ denote the weighting factors associated with speed, flux, and torque, respectively. Similarly to the previous examples, a control objective i_{oc}, defined mathematically in Eq. (8), is included in Eq. (14) to prevent excessive currents in the induction machine.

3.3.2 Rotor flux and load torque estimation

To implement the MPDSC scheme of an induction machine, flux and load torque information is needed, and these variables are estimated here through an EKF observer using the measured speeds, voltages, and currents. To this end, the induction machine model in Eq. (1) is extended to include load torque τ_l, considering the viscous friction term $B_t\omega_m$ as a part of the τ_l. The new state vector $\boldsymbol{x}_t$ and nonlinear system model $\boldsymbol{f}(\boldsymbol{x}_t, \boldsymbol{u}_t)$ are as follows:

$$x_t = \begin{bmatrix} i_{s\alpha} & i_{s\beta} & \omega_m & \psi_{r\alpha} & \psi_{r\beta} & \tau_l \end{bmatrix}^T$$

and

$$f(x_t, u_t) = \begin{bmatrix} -\frac{1}{T_\sigma} i_{s\alpha} + \frac{k_r}{T_r L_\sigma} \psi_{r\alpha} + \frac{k_r}{L_\sigma} \omega_r \psi_{r\beta} + \frac{1}{L_\sigma} u_{s\alpha} \\ -\frac{1}{T_\sigma} i_{s\beta} + \frac{k_r}{L_\sigma T_r} \psi_{r\beta} - \frac{k_r}{L_\sigma} \omega_r \psi_{r\alpha} + \frac{1}{L_\sigma} u_{s\beta} \\ \frac{3}{2} k_r \frac{p_p}{J_t} \left(\psi_{r\alpha} i_{s\beta} - \psi_{r\beta} i_{s\alpha} \right) - \frac{1}{J_t} \tau_l \\ \frac{L_m}{T_r} i_{s\alpha} - \frac{1}{T_r} \psi_{r\alpha} - \omega_r \psi_{r\beta} \\ \frac{L_m}{T_r} i_{s\beta} - \frac{1}{T_r} \psi_{r\beta} + \omega_r \psi_{r\alpha} \\ 0 \end{bmatrix}$$

The extended continuous-time model given above should be represented in discrete form. To achieve this, the first-order Euler approximation in Eq. (2) is utilized here. With the above discrete model and the recursions outlined in **Table 1**, an EKF can now be constructed and integrated into the classic MPC schemes.

This obviously means that MPDSC has higher computational complexity than MPCC and MPTC. However, it can be ignored in applications where higher control performance is required.

The resulting control performance is shown in **Figure 7** for a sampling time of $T = 25\ \mu s$. It is observed a good performance of the observer because it can accurately estimate the real states and maintain the induction machine states on the reference trajectories with a good performance of the currents, the flux, the torque and the speed to their respective references with a fast dynamic response.

Definitions:	$F_{k+1\vert k} = \left.\frac{\partial f(x, u_k)}{\partial x}\right\vert_{x=x_k}$ $H_k = \left.\frac{\partial h(x)}{\partial x}\right\vert_{x=x_k^-}$
Initialization: For $k = 0$, set	$\hat{x}_0 = E[x_0]$ $P_0 = E\left[(x_0 - E[x_0])(x_0 - E[x_0])^T\right]$
For $k = 1, 2, \ldots$, compute: *Time update:*	$\hat{x}_k^- = f(\hat{x}_{k-1}, u_k)$ $P_k^- = F_{k\vert k-1} P_{k-1} F_{k\vert k-1}^T + Q_{k-1}$
Measurement update:	$K_k = P_k^- H^T \left[H P_k^- H^T + R_k \right]^{-1}$ $\hat{x}_k = \hat{x}_k^- + K_k (z_k - H \hat{x}_k^-)$ $P_k = (I - K_k H) P_k^-$

$F_{k+1|k}$ is the transition matrix, H is the measurement matrix, Q_k is the system noise covariance matrix, K_k is the Kalman gain, R_k is the measurement noise covariance matrix, P_k^- and P_k are the priori and posteriori covariance matrices, respectively.

Table 1.
Extended Kalman Filter (EKF).

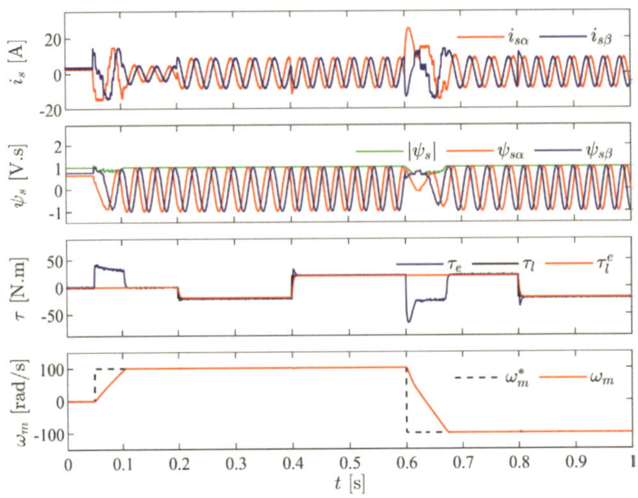

Figure 7.
Control performance of the MPDSC scheme $(T = 25\ \mu s)$.

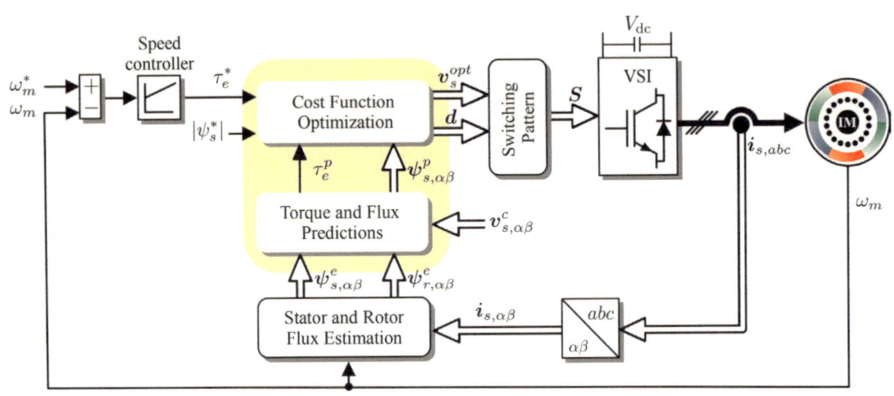

Figure 8.
Block diagram of the modulated MPTC scheme.

4. Modulated FS-MPC

Various modulated MPC schemes have been proposed to reduce current harmonics and torque ripples. This section introduces the application of two modulated MPC schemes through the MPTC scheme and compares them with the conventional MPTC scheme with a sampling time of 100 μs. Modulated MPC schemes, depicted in **Figure 8**, have steps similar to conventional MPC schemes, that is, prediction and cost function optimization phases, apart from the modulation phase. They use two active VVs and one zero VV at each sampling interval, unlike conventional MPC schemes that use only a single VV. Different approaches are used for selecting these two active VVs: one is based on using adjacent VVs and the other is based on the use of VVs with minimum cost values.

4.1 Modulated MPTC based on adjacent VVs

When two adjacent active VVs ($\boldsymbol{v}_1$ and $\boldsymbol{v}_2$) and one zero VV ($\boldsymbol{v}_0$) are used, the duty cycle of each vector is calculated based on their corresponding cost values as follows:

$$d_i = K/g_i \tag{15}$$

$$\sum_{i=0}^{2}(d_i) = 1 \tag{16}$$

where d_i and g_i are the corresponding duty cycle and cost value of $\boldsymbol{v}_i, i \in \{0,1,2\}$, respectively. $\{\boldsymbol{v}_1, \boldsymbol{v}_2\} \in \{\{1,2\},\{2,3\},\{3,4\},\{4,5\},\{5,6\},\{6,1\}\}$.

Based on these expressions, the parameter K and duty cycles for each VV are calculated as follows:

$$K = \frac{g_0 g_1 g_2}{g_0 g_1 + g_1 g_2 + g_0 g_2} \tag{17}$$

$$d_0 = \frac{T_s g_1 g_2}{g_0 g_1 + g_1 g_2 + g_0 g_2} \tag{18}$$

$$d_1 = \frac{T_s g_0 g_2}{g_0 g_1 + g_1 g_2 + g_0 g_2} \tag{19}$$

$$d_2 = \frac{T_s g_0 g_1}{g_0 g_1 + g_1 g_2 + g_0 g_2} \tag{20}$$

The final cost function can be defined as follows, based on the cost values and duty cycles of the corresponding VVs:

$$g = d_1 g_1 + d_2 g_2 \tag{21}$$

To apply the optimal voltage vectors and the zero vector, the time periods T_0, T_1, and T_2 are calculated using the duty cycles d_0, d_1, and d_2 as follows:

$$T_i = \frac{T_s d_i}{t_m}, \tag{22}$$

$$T_s = \sum_{i=0}^{2}(T_i) \tag{23}$$

where t_m corresponds to the division of the sampling time T_s into small equal parts where the optimal vectors will be applied. Once the optimal vectors $\boldsymbol{v}_1$ and $\boldsymbol{v}_2$ and the times at which they will be applied (T_0, T_1, and T_2) have been obtained, a commutation strategy consisting of seven steps is established, as shown in **Figure 9**.

4.2 Modulated MPTC based on the VVs with minimum cost values

As an alternative to adjacent active VVs, active VVs with minimum cost values can be used. For this purpose, the cost value of each VV is first calculated using prediction models and the cost function. These values are then sorted, and the VVs with the

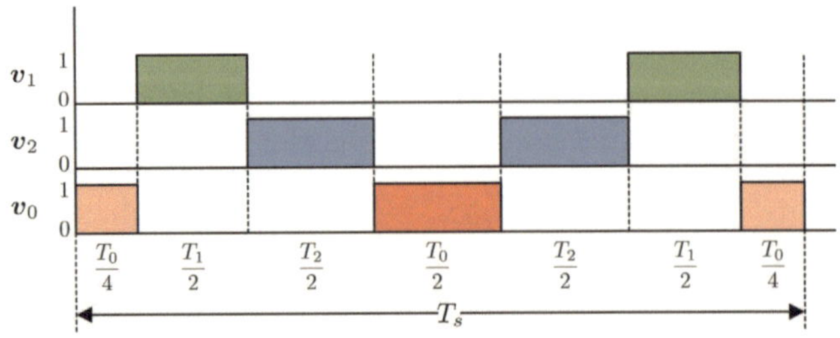

Figure 9.
Application of VVs for single sampling intervals in modulated MPC schemes.

lowest cost values are selected. Finally, the duties and time periods are calculated based on Eqs. (18)–(20), (22) and (23), respectively.

The comparative results between conventional and modulated MPC schemes are presented in **Figure 10**. The results indicate that modulated MPCs significantly reduce the current harmonics and torque ripples.

5. MPC schemes without weighting factors

Some conventional MPC schemes, particularly MPTC and MPDSC, require the use of weighting factors in their cost functions due to the various error terms with different units. In this way, the effect of each term on the selection of optimal VV can be balanced. However, the challenge of determining optimal weighting factors, combined with changes in operating conditions, complicates the design of MPC schemes, and improper selection of these factors can directly impact their control performance. To address this issue, various MPC schemes without weighting factors have been proposed by Zerdali et al. [17].

This section introduces two weighting factor-free MPC schemes: the sequential MPTC scheme and the decision-making-based MPTC scheme. Since these techniques evaluate each term in the cost function separately, the first step is to define each term as a separate cost function, as follows:

$$g_\tau = \left|\tau_e^* - \tau_{e,k+1}^p\right| \tag{24}$$

$$g_\psi = \left|\left|\boldsymbol{\psi}_s^*\right| - \left|\boldsymbol{\psi}_{s,k+1}^p\right|\right| \tag{25}$$

5.1 Sequential MPTC scheme

Sequential MPC is one of the most popular MPC schemes that eliminates weighting factors. This section introduces this technique through the MPTC scheme, as illustrated in **Figure 11**. The resulting control performance is presented in **Figure 12**.

This technique evaluates the above cost functions sequentially. In the first stage, all possible VVs are considered for the first cost function, and a few VVs with the lowest torque errors are selected for the next stage. In the second stage, the selected VVs that minimize the first cost function are then evaluated against the second cost function.

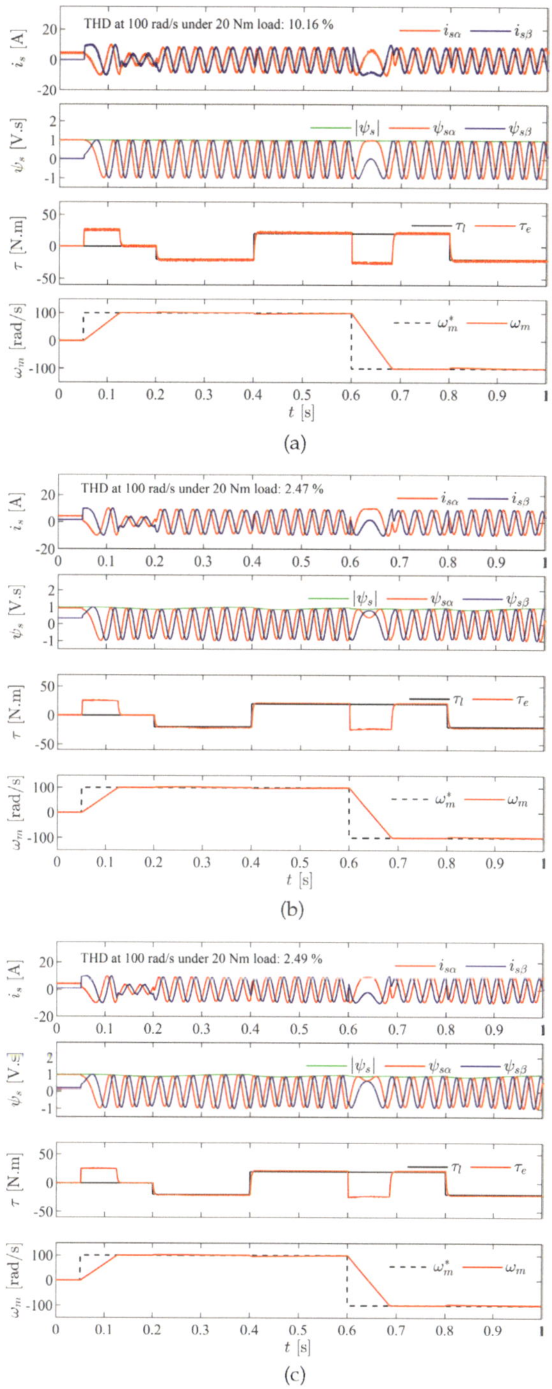

Figure 10.
Comparison of conventional and modulated MPTC schemes: (a) conventional MPTC scheme and modulated MPTC schemes with (b) a zero VV and two adjacent active VVs and (c) a zero VV and two active VVs having the lowest cost values ($T = 100\ \mu s$ for all).

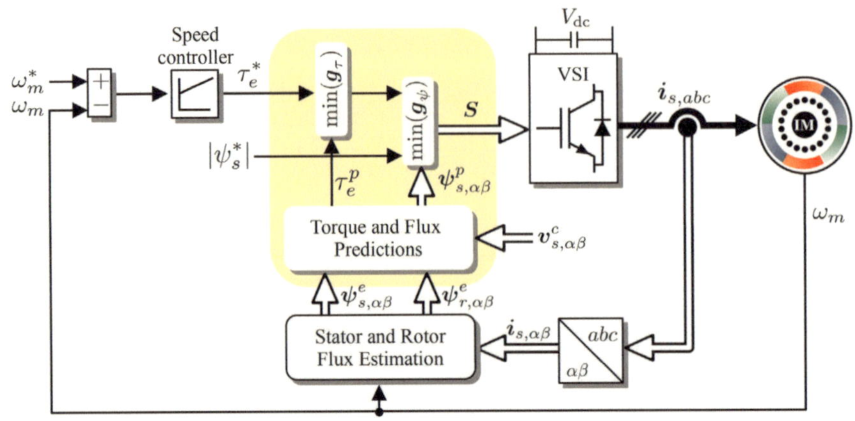

Figure 11.
Block diagram of sequential MPTC scheme.

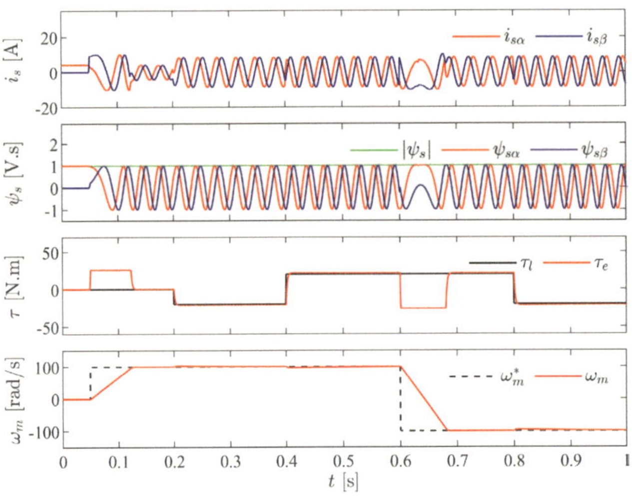

Figure 12.
Control performance of sequential MPTC scheme ($T = 25\ \mu s$).

As a result, a single VV is determined at the end of the optimization phase (**Figure 12**).

5.2 Decision-making-based MPTC scheme

Various decision-making-based MPC schemes have been proposed to eliminate the weighting factors. However, decision-making based on the Euclidean norm offers lower design and computational complexity, as it does not require a sorting algorithm [18]. This section presents this technique within the MPTC scheme, as illustrated in **Figure 13**. It follows a prediction phase similar to that of the conventional MPTC but differs in cost function optimization.

The following steps illustrate the details of this technique in turn:

First step: Generating the data set by evaluating each VV for each cost function.

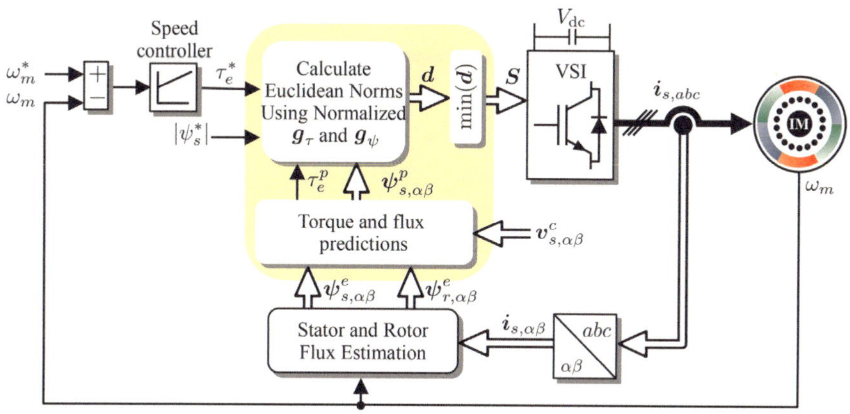

Figure 13.
Block diagram of decision-making-based MPTC scheme.

$$X_{ij} = \begin{pmatrix} g_\tau(v_0) & g_\psi(v_0) \\ g_\tau(v_1) & g_\psi(v_1) \\ \vdots & \vdots \\ g_\tau(v_7) & g_\psi(v_7) \end{pmatrix} \tag{26}$$

where i and j denote the voltage vector and cost function, respectively.
Second step: Normalization of the dataset.

$$Y_{ij} = \frac{\min(x_j) - x_{ij}}{\min(x_j) - \max(x_j)} \tag{27}$$

where x_j indicates the column vectors in the dataset.
Third step: Computing the Euclidean norm over the normalized dataset Y_{ij}.

$$d_i = \|y_i\|_2 \tag{28}$$

y_i denotes the row vectors in the normalized dataset.
Fourth step: The selection of VV that minimizes the Euclidean norm.

$$v_{\text{opt}} = \arg\min_i d_i \tag{29}$$

It is clear that by completing the above steps, the optimum VV is determined without the need for weighting factors. The corresponding control performance is presented in **Figure 14**.

6. Sensorless control

Sensorless control enhances reliability against sensor failures and increases robustness to noise, while also reducing cost, complexity, and size of electric drives by eliminating the need for extra cabling and hardware. In certain applications, it may not even be possible to install a sensor for measurements, necessitating the use of

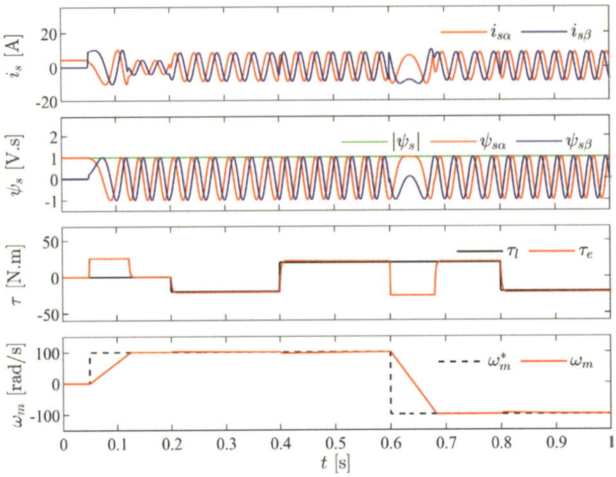

Figure 14.
Control performance of decision-making-based MPTC scheme ($T = 25\ \mu s$).

sensorless control. Various methods have been proposed to leverage both FCS-MPC and sensorless control. To illustrate some applications of sensorless MPC schemes, this section applies sensorless control to the classic MPC schemes discussed earlier. An EKF observer is first introduced to estimate flux, mechanical speed, and load torque and is then integrated with the MPC schemes.

6.1 Flux, speed, and load torque estimation

To estimate rotor flux, mechanical speed, and load torque, the function $\mathbf{f}(\mathbf{x}_t, \mathbf{u}_t)$ in Eq. (1) is extended by load torque in addition to an update in the function $\mathbf{h}(\mathbf{x}_t)$. In this case, the corresponding vectors are as follows:

$$\mathbf{x}_t = \begin{bmatrix} i_{s\alpha} & i_{s\beta} & \psi_{r\alpha} & \psi_{r\beta} & \omega_m & \tau_l \end{bmatrix}^T$$
$$\mathbf{u}_t = \begin{bmatrix} u_{s\alpha} & u_{s\beta} \end{bmatrix}^T$$
$$\mathbf{h}(\mathbf{x}_t) = \begin{bmatrix} i_{s\alpha} & i_{s\beta} \end{bmatrix}^T$$

and

$$\mathbf{f}(\mathbf{x}_t, \mathbf{u}_t) = \begin{bmatrix} -\frac{1}{T_\sigma} i_{s\alpha} + \frac{k_r}{T_r L_\sigma} \psi_{r\alpha} + \frac{k_r}{L_\sigma} \omega_r \psi_{r\beta} + \frac{1}{L_\sigma} u_{s\alpha} \\ -\frac{1}{T_\sigma} i_{s\beta} + \frac{k_r}{L_\sigma T_r} \psi_{r\beta} - \frac{k_r}{L_\sigma} \omega_r \psi_{r\alpha} + \frac{1}{L_\sigma} u_{s\beta} \\ \frac{L_m}{T_r} i_{s\alpha} - \frac{1}{T_r} \psi_{r\alpha} - \omega_r \psi_{r\beta} \\ \frac{L_m}{T_r} i_{s\beta} - \frac{1}{T_r} \psi_{r\beta} + \omega_r \psi_{r\alpha} \\ \frac{3}{2} k_r \frac{p_p}{J_t} \left(\psi_{r\alpha} i_{s\beta} - \psi_{r\beta} i_{s\alpha} \right) - \frac{1}{J_t} \tau_l \\ 0 \end{bmatrix}$$

Applying the first-order Euler approximation in Eq. (2) to the above model and using the discrete model in the recursions of the EKF observer as outlined in **Table 1**, an EKF observer is obtained for estimating rotor flux, rotor mechanical speed, and load torque [19].

6.2 Sensorless MPCC scheme

A sensorless MPCC scheme, presented in **Figure 15**, is obtained by integrating the EKF observer designed in Section 6.1 into the classic MPCC scheme in Section 3.1. It is worth noting that load torque estimation is not necessary for the sensorless MPCC scheme; it is only used to improve the low-speed estimation performance of the EKF observer [20].

As shown in **Figure 15**, the variables filtered by the EKF are used in the prediction model instead of the measured ones. This approach helps reduce noise in the predictions and improves control performance.

The resulting control performance is shown in **Figure 16** for a sampling time of $T = 25\ \mu s$.

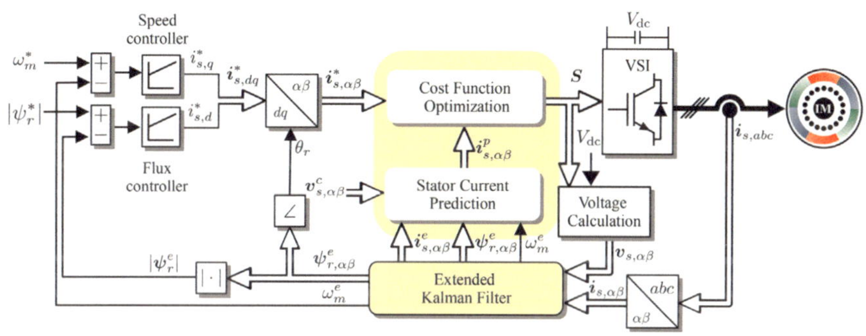

Figure 15.
Block diagram of the sensorless MPCC scheme.

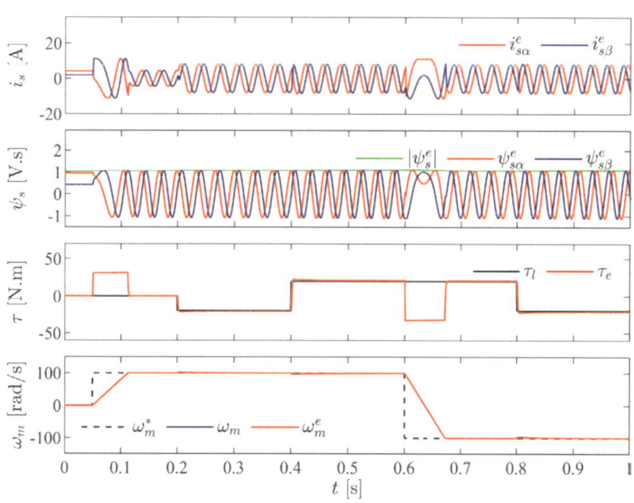

Figure 16.
Control performance of sensorless MPCC scheme ($T = 25\ \mu s$).

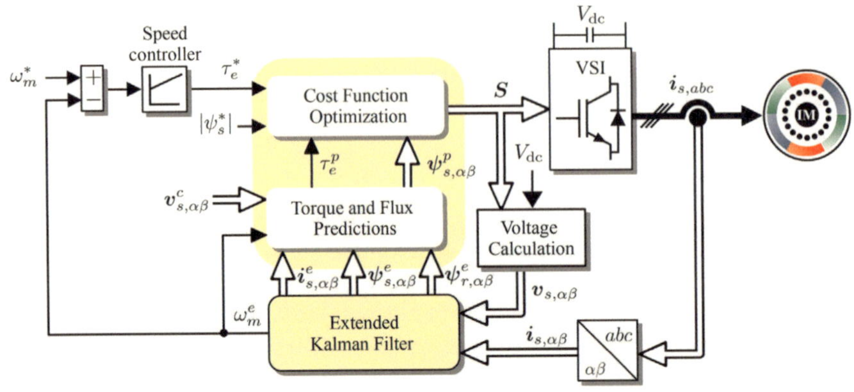

Figure 17.
Block diagram of the sensorless MPTC scheme.

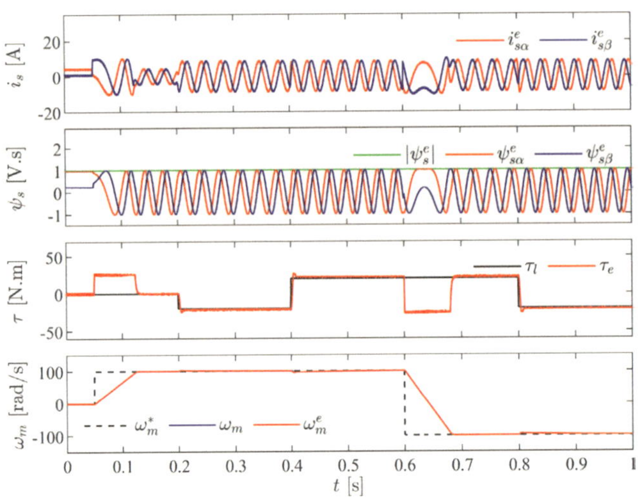

Figure 18.
Control performance of sensorless MPTC scheme ($T = 25\ \mu s$).

6.3 Sensorless MPTC scheme

A sensorless MPTC scheme, illustrated in **Figure 17**, is designed by combining the MPTC scheme in Section 3.2 and the EKF observer in Section 6.1. Similar to the sensorless MPCC scheme, the filtered states from the EKF observer are used in the prediction models rather than the measured ones. This approach helps to mitigate the adverse effects of noisy measurements. The resulting control performance is shown in **Figure 18** for a sampling time of $T = 25\ \mu s$.

6.4 Sensorless MPDSC scheme

A sensorless MPDSC scheme, shown in **Figure 19**, is designed by combining the MPDSC scheme in Section 3.3 and the EKF observer in Section 6.1. Unlike the conventional MPDSC scheme, the mechanical speed and load torque, which are essential for prediction models and cost function optimization, are estimated by the EKF

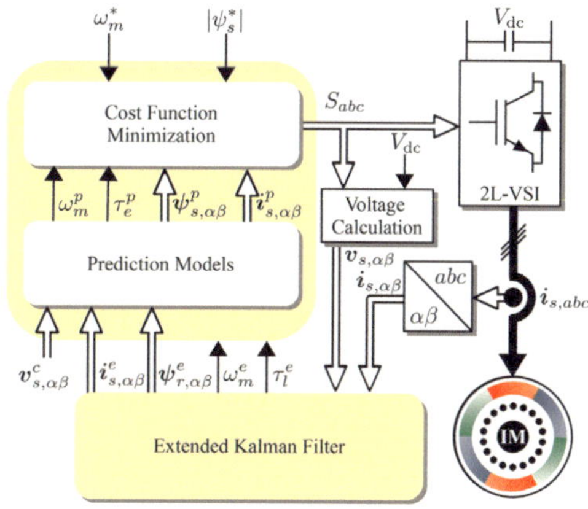

Figure 19.
Block diagram of the sensorless MPDSC scheme.

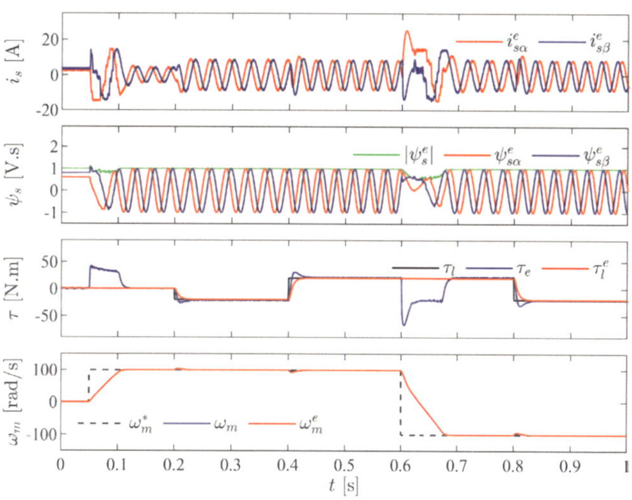

Figure 20.
Control performance of sensorless MPDSC scheme ($T = 25\ \mu s$).

observer, as shown in **Figure 19**. Similar to previous sensorless direct MPC schemes, the filtered states from the EKF observer are utilized in the prediction models to suppress the adverse effects of noise on the predictions. The resulting control performance is presented in **Figure 20**.

7. Summary

This chapter first introduced conventional FCS-MPC schemes for controlling an induction motor fed by a two-level VSI. Then, two modulated schemes were applied to conventional MPCs to reduce current harmonics and torque ripples. To address the issue of weighting factor design in the FCS-MPC technique, two weighting factor elimination methods were introduced using the conventional MPTC scheme. Lastly, a sensorless control technique was integrated with conventional FCS-MPC schemes to reduce cost and drive size while enhancing reliability and robustness.

All these applications demonstrate that MPC is an effective method for controlling power converters and electric drives. Its flexibility allows it to address challenges with minimal effort. However, it is clear that further research is still needed to resolve issues with even less effort.

Author details

Marco Rivera[1*†], Emrah Zerdali[2†], Patrick Wheeler[1†] and José Rodríguez[3†]

1 Department of Electrical and Electronics Engineering, Faculty of Engineering, Nottingham, UK

2 Department of Electrical and Electronics Engineering, Faculty of Engineering, Ege University, Izmir, Turkey

3 Faculty of Engineering, Universidad San Sebastián, Santiago, Chile

*Address all correspondence to: marco.rivera@nottingham.ac.uk

† These authors contributed equally.

References

[1] Becker M, Stender M, Wallscheid O. Nonlinear Efficiency-Optimal Model Predictive Torque Control of Induction Machines. IEEE Journal of Emerging and Selected Topics in Power Electronics. Oct. 2024;**12**(5):4740-4753. DOI: 10.1109/JESTPE.2024.3437739

[2] Tang Y, Xu W, Ge J, Shangguan Y, Li Y, Yuan W. Predictive flux control-based efficiency optimization strategy for three-level inverter-fed linear induction machine adapted to linear metro. IEEE Transactions on Power Electronics. 2024;**39**(10):13672-13685. DOI: 10.1109/TPEL.2024.3425829

[3] Wang S, Zhang Y, Zhang Q, Zhao J, Zhang Q. MPTC for dual three-phase PMSM based on novel virtual voltage vector. IEEE Transactions on Energy Conversion. 2024;**39**(3):1488-1496. DOI: 10.1109/TEC.2024.3366182

[4] Zhang Z, Chen J, Han R, Wu Y, Gong Y, Chang S. K-best-sphere-decoding-based model predictive control for dual three-phase SPMSMs. IEEE Transactions on Magnetics. 2024;**60**(9): 1-6. DOI: 10.1109/TMAG.2024. 3406448

[5] Xu W, Xiao H, Ge J, Tang Y, Li Y, Junejo AK, et al. Discrete space vector modulation-based MPCC with inductance identification for thrust ripple suppression in linear induction machine. IEEE Transactions on Power Electronics. 2024;**39**(10):12883-12893. DOI: 10.1109/TPEL.2024.3425726

[6] Aboelhassan A, Wang S, Mikhaylov Y, Buticchi G, Galea M, Bozhko S. Discontinuous modulated model predictive control for low inductance high-speed electric drive applications. IEEE Transactions on Industrial Electronics. 2024;**71**(7): 6689-6698. DOI: 10.1109/TIE.2023.3301532

[7] Hamad SA, Xu W, Junejo AK, Xiao H, Liao K, Tang Y, et al. Improved full order sliding mode based on model predictive flux control for LIM applied to urban transit. IEEE Transactions on Transportation Electrification. 2024:1-1. DOI: 10.1109/TTE.2024.3424496

[8] Jia Z, Song W, Ma C, Zhang B, Sun N. An MPC-based online inter-turn fault diagnosis method for induction motors with fault localization. IEEE Transactions on Transportation Electrification. 2024:1-1. DOI: 10.1109/TTE.2024.3386847

[9] Chakraborty R, Das S. Fault-tolerant technique against current sensors for model predictive control of induction motor drives. IEEE Transactions on Power Electronics. 2024;**39**(7): 8506-8516. DOI: 10.1109/TPEL.2024.3381169

[10] Laadjal K, Bento F, Serra J, Cardoso AJM. An integrated strategy for the real-time detection and discrimination of stator inter-turn short-circuits and converter faults in asymmetrical six-phase induction motors. IEEE Transactions on Industry Applications. 2024;60(1):377-387. doi: 10.1109/TIA.2023.3321266

[11] Manikandan R, Singh RR. Open circuit fault localization in NPC inverter fed induction motor drives using cost function and estimated voltages. IEEE Transactions on Circuits and Systems II: Express Briefs. 2024;**71**(2):922-926. DOI: 10.1109/TCSII.2023.3310671

[12] Su Z, Sun X, Lei G, Yao M. Model-free predictive current control for dual

three-phase PMSM drives with an optimal modulation pattern. IEEE Transactions on Industrial Electronics. 2024;**71**(9):10140-10149. DOI: 10.1109/TIE.2023.3327554

[13] Gonzalez-Prieto A, Gonzalez-Prieto I, Dordevic O, Aciego JJ, Montenegro J, Duran MJ, et al. Memory-based model predictive control for parameter detuning in multiphase electric machines. IEEE Transactions on Power Electronics. 2024;**39**(2):2546-2557. DOI: 10.1109/TPEL.2023.3328427

[14] Trivedi C, Bhattacharya P, Prasad VK, Patel V, Singh A, Tanwar S, et al. Explainable AI for industry 5.0: Vision, architecture, and potential directions. IEEE Open Journal of Industry Applications. 2024;**5**:177-208. DOI: 10.1109/OJIA.2024.3399057

[15] Boyd JD, Reising DR, Murphy AM, Kuhlers JD, McAmis CM, Rossman JB. Machine learning techniques to predict voltage unbalance in a power transmission system. IEEE Open Journal of Industry Applications. 2024;**5**:86-93. DOI: 10.1109/OJIA.2024.3369993

[16] Yang L, Li H, Huang J, Zhang Z, Zhao H. Model predictive direct speed control with novel cost function for SMPMSM drives. IEEE Transactions on Power Electronics. 2022;**37**(8):9586-9595. DOI: 10.1109/TPEL.2022.3155465

[17] Zerdali E, Rivera M, Wheeler P. A review on weighting factor design of finite control set model predictive control strategies for ac electric drives. IEEE Transactions on Power Electronics. 2024;**39**(8):9967-9981. DOI: 10.1109/TPEL.2024.3370550

[18] Zerdali E, Altintas M, Bakbak A, Mese E. Computationally efficient predictive torque control strategies without weighting factors. Turkish Journal of Electrical Engineering and Computer Sciences. 2022;**30**:2554-2567. DOI: 10.55730/1300-0632.3955

[19] Yildiz R, Barut M, Zerdali E. A comprehensive comparison of extended and unscented Kalman filters for speed-sensorless control applications of induction motors. IEEE Transactions on Industrial Informatics. 2020;**16**(10):6423-6432. DOI: 10.1109/TII.2020.2964876

[20] Zerdali E, Yildiz R, Inan R, Demir R, Barut M. Improved speed and load torque estimations with adaptive fading extended Kalman filter. International Transactions on Electrical Energy Systems. 2021;**31**(1):e12684. DOI: 10.1002/2050-7038.12684